PRESS KIT

APOLLO 11
LUNAR LANDING MISSION

NASA

NATIONAL AERONAUTICS AND SPACE ADMINISTRATION

Cover: Apollo 11 mission commander Neil Armstrong took this iconic photograph of Buzz Aldrin walking on the moon.

Published by Books Express Publishing
Copyright © Books Express, 2012
ISBN 978-1-78039-860-0

Books Express publications are available from all good retail and online booksellers. For publishing proposals and direct ordering please contact us at: info@books-express.com

NATIONAL AERONAUTICS AND SPACE ADMINISTRATION
WASHINGTON, D.C. 20546

TELS. WO 2-4155
WO 3-6925

FOR RELEASE: SUNDAY
July 6, 1969

RELEASE NO: 69-83K

PROJECT: APOLLO 11

(To be launched no
earlier than July 16)

contents

GENERAL RELEASE--1-17
APOLLO 11 COUNTDOWN---18-20
LAUNCH EVENTS---21
APOLLO 11 MISSION EVENTS--22-25
MISSION TRAJECTORY AND MANEUVER DESCRIPTION---------------------------26
   Launch--26-30
   Earth Parking Orbit (EPO)--30
   Translunar Injection (TLI)---30
   Transposition, Docking and Ejection (TD&E)----------------------30-32
   Translunar Coast--33
   Lunar Orbit Insertion (LOI)---------------------------------------33
   Lunar Module Descent, Lunar Landing---------------------------33-41
   Lunar Surface Extravehicular Activity (EVA)-------------------42-47
   Lunar Sample Collection---48
   LM Ascent, Lunar Orbit Rendezvous-----------------------------49-53
   Transearth Injection (TEI)------------------------------------53-56
   Transearth Coast--57
   Entry Landing---57-63
RECOVERY OPERATIONS, QUARANTINE-------------------------------------64-65
   Lunar Receiving Laboratory------------------------------------65-67
SCHEDULE FOR TRANSPORT OF SAMPLES, SPACECRAFT & CREW-----------------68
LUNAR RECEIVING LABORATORY PROCEDURES TIMELINE
   (TENTATIVE)---69-70
APOLLO 11 GO/NO-GO DECISION POINTS----------------------------------71
APOLLO 11 ALTERNATE MISSIONS-------------------------------------72-73
ABORT MODES---74
   Deep Space Aborts--74-76
ONBOARD TELEVISION--77
   Tentative Apollo 11 TV Times--------------------------------------78
PHOTOGRAPHIC TASKS---79-80
LUNAR DESCRIPTION---81
   Physical Facts--81
   Apollo Lunar Landing Sites-----------------------------------82-85

Contents Continued   2

COMMAND AND SERVICE MODULE STRUCTURE, SYSTEMS---------------86-88
    CSM Systems--88-95
LUNAR MODULE STRUCTURES, WEIGHT-----------------------------96
    Ascent Stage---96-101
    Descent Stage--101-103
    Lunar Module Systems---------------------------------103-107
SATURN V LAUNCH VEHICLE DESCRIPTION & OPERATION-------------108
    Launch Vehicle Range Safety Provisions---------------108-109
    Space Vehicle Weight Summary-------------------------110-111
    First Stage--112
    Second Stage---112-113
    Third Stage--113
    Instrument Unit--------------------------------------113-114
    Propulsion---114-115
    Launch Vehicle Instrumentation and Communication-----115
    S-IVB Restart--116
    Differences in Launch Vehicles for A-10 and A-11-----116
APOLLO 11 CREW--117
    Life Support Equipment - Space Suits-----------------117-122
    Apollo 11 Crew Menu----------------------------------123-132
    Personal Hygiene-------------------------------------133
    Medical Kit--133
    Survival Gear--133-135
    Biomedical Inflight Monitoring-----------------------135
    Training---136-137
    Crew Biographies-------------------------------------138-144
EARLY APOLLO SCIENTIFIC EXPERIMENTS PACKAGE----------------145-153
APOLLO LUNAR RADIOISOTOPIC HEATER (ALRH)-------------------154-157
APOLLO LAUNCH OPERATIONS-----------------------------------158
    Prelaunch Preparations-------------------------------158-160
LAUNCH COMPLEX 39--161
    Vehicle Assembly Building----------------------------162-163
    Launch Control Center--------------------------------163-164
    Mobile Launcher--------------------------------------164-165
    Transporter--165-166
    Crawlerway---166
    Mobile Service Structure-----------------------------166-167
    Water Deluge System----------------------------------167
    Flame Trench and Deflector---------------------------167-168
    Pad Areas--168
    Mission Control Center-------------------------------169-170
MANNED SPACE FLIGHT NETWORK--------------------------------171-174
    NASA Communications Network--------------------------174-176
    Network Computers------------------------------------176-177
    The Apollo Ships-------------------------------------178
    Apollo Range Instrumentation Aircraft (ARIA)---------179
    Ship Positions for Apollo 11-------------------------180

CONTAMINATION CONTROL PROGRAM-------------------------------------181
    Lunar Module Operations------------------------------------181-187
    Command Module Operations----------------------------------187
    Lunar Mission Recovery Operations--------------------------187-188
    Biological Isolation Garment-------------------------------188
    Mobile Quarantine Facility---------------------------------188
    Lunar Receiving Laboratory---------------------------------189-190
    Sterilization and Release of Spacecraft-------------------190-191
APOLLO PROGRAM MANAGEMENT---192
    Apollo/Saturn Officials------------------------------------193-217
    Major Apollo/Saturn V Contractors--------------------------218-219
PRINCIPAL INVESTIGATORS AND INVESTIGATIONS OF
    LUNAR SURFACE SAMPLES--------------------------------------220-241
APOLLO GLOSSARY--242-246
APOLLO ACRONYMS AND ABBREVIATIONS----------------------------------247-248
CONVERSION FACTORS---249-250

- 0 -

NEWS

NATIONAL AERONAUTICS AND SPACE ADMINISTRATION
WASHINGTON, D.C. 20546

TELS. WO 2-4155
WO 3-6925

FOR RELEASE: SUNDAY
July 6, 1969

RELEASE NO: 69-83K

APOLLO 11

The United States will launch a three-man spacecraft toward the Moon on July 16 with the goal of landing two astronaut-explorers on the lunar surface four days later.

If the mission--called Apollo 11--is successful, man will accomplish his long-time dream of walking on another celestial body.

The first astronaut on the Moon's surface will be 38-year-old Neil A. Armstrong of Wapakoneta, Ohio, and his initial act will be to unveil a plaque whose message symbolizes the nature of the journey.

Affixed to the leg of the lunar landing vehicle, the plaque is signed by President Nixon, Armstrong and his Apollo 11 companions, Michael Collins and Edwin E. Aldrin, Jr.

-more-   6/26/69

It bears a map of the Earth and this inscription:

HERE MEN FROM THE PLANET EARTH

FIRST SET FOOT UPON THE MOON

JULY 1969 A.D.

WE CAME IN PEACE FOR ALL MANKIND

The plaque is fastened to the descent stage of the lunar module and thus becomes a permanent artifact on the lunar surface.

Later Armstrong and Aldrin will emplant an American flag on the surface of the Moon.

The Apollo 11 crew will also carry to the Moon and return two large American flags, flags of the 50 states, District of Columbia and U.S. Territories, flags of other nations and that of the United Nations Organization.

During their 22-hour stay on the lunar surface, Armstrong and Aldrin will spend up to 2 hours and 40 minutes outside the lunar module, also gathering samples of lunar surface material and deploying scientific experiments which will transmit back to Earth valuable data on the lunar environment.

Apollo 11 is scheduled for launch at 9:32 a.m. EDT July 16 from the National Aeronautics and Space Administration's Kennedy Space Center Launch Complex 39-A. The mission will be the fifth manned Apollo flight and the third to the Moon.

-more-

The prime mission objective of Apollo 11 is stated simply: "Perform a manned lunar landing and return". Successful fulfillment of this objective will meet a national goal of this decade, as set by President Kennedy May 25, 1961.

Apollo 11 Commander Armstrong and Command Module Pilot Collins 38, and Lunar Module Pilot Aldrin, 39, will each be making his second space flight. Armstrong was Gemini 8 commander, and backup Apollo 8 commander; Collins was Gemini 10 pilot and was command module pilot on the Apollo 8 crew until spinal surgery forced him to leave the crew for recuperation; and Aldrin was Gemini 12 pilot and Apollo 8 backup lunar module pilot. Armstrong is a civilian, Collins a USAF lieutenant colonel and Aldrin a USAF colonel.

Apollo 11 backup crewmen are Commander James A. Lovell, Command Module Pilot William A. Anders, both of whom were on the Apollo 8 first lunar orbit mission crew, and Lunar Module Pilot Fred W. Haise.

The backup crew functions in three significant categories. They help the prime crew with mission preparation and hardware checkout activities. They receive nearly complete mission training which becomes a valuable foundation for later assignment as a prime crew and finally, should the prime crew become unavailable, they are prepared to fly as prime crew on schedule up until the last few weeks at which time full duplicate training becomes too costly and time consuming to be practical.

Apollo 11, after launch from Launch Complex 39-A, will begin the three-day voyage to the Moon about two and a half hours after the spacecraft is inserted into a 100-nautical mile circular Earth parking orbit. The Saturn V launch vehicle third stage will restart to inject Apollo 11 into a translunar trajectory as the vehicle passes over the Pacific midway through the second Earth parking orbit.

The "go" for translunar injection will follow a complete checkout of the space vehicle's readiness to be committed for injection. About a half hour after translunar injection (TLI), the command/service module will separate from the Saturn third stage, turn around and dock with the lunar module nested in the spacecraft LM adapter. Spring-loaded lunar module holddowns will be released to eject the docked spacecraft from the adapter.

APOLLO 11 —— Launch And Translunar Injection

Check Of Systems

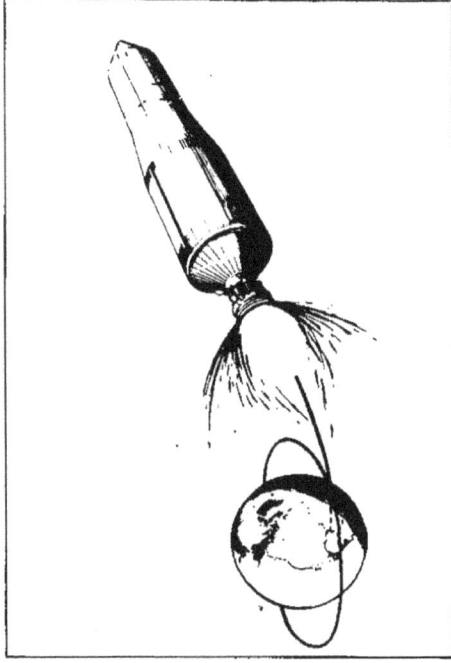

Translunar Injection

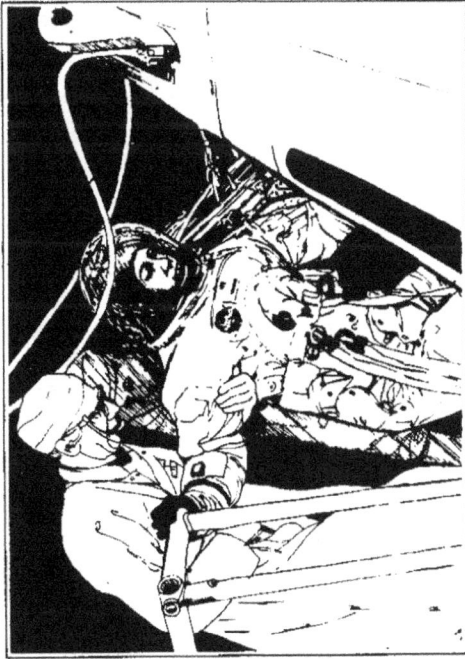

Astronaut Insertion

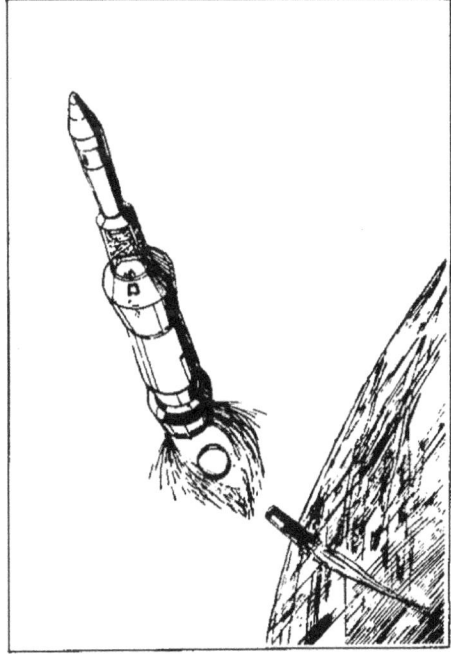

Saturn Staging

Later, leftover liquid propellant in the Saturn third stage will be vented through the engine bell to place the stage into a "slingshot" trajectory to miss the Moon and go into solar orbit.

During the translunar coast, Apollo 11 will be in the passive thermal control mode in which the spacecraft rotates slowly about one of its axes to stabilize thermal response to solar heating. Four midcourse correction maneuvers are possible during translunar coast and will be planned in real time to adjust the trajectory.

Apollo 11 will first be inserted into a 60-by-170-nautical mile elliptical lunar orbit, which two revolutions later will be adjusted to a near-circular 54 x 66 nm. Both lunar orbit insertion burns (LOI), using the spacecraft's 20,500-pound-thrust service propulsion system, will be made when Apollo 11 is behind the Moon and out of "sight" of Manned Space Flight Network stations.

Some 21 hours after entering lunar orbit, Armstrong and Aldrin will man and check out the lunar module for the descent to the surface. The LM descent propulsion system will place the LM in an elliptical orbit with a pericynthion, or low point above the Moon, of 50,000 feet, from which the actual descent and touchdown will be made.

APOLLO 11 — Translunar Flight

Extraction Of Lunar Module

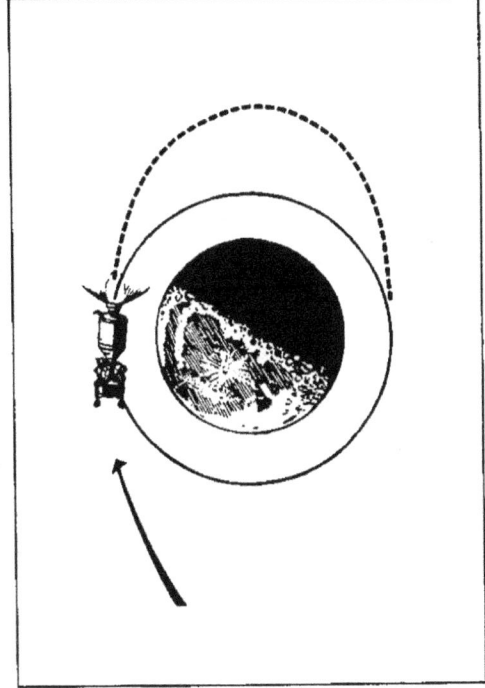

Lunar Orbit Insertion

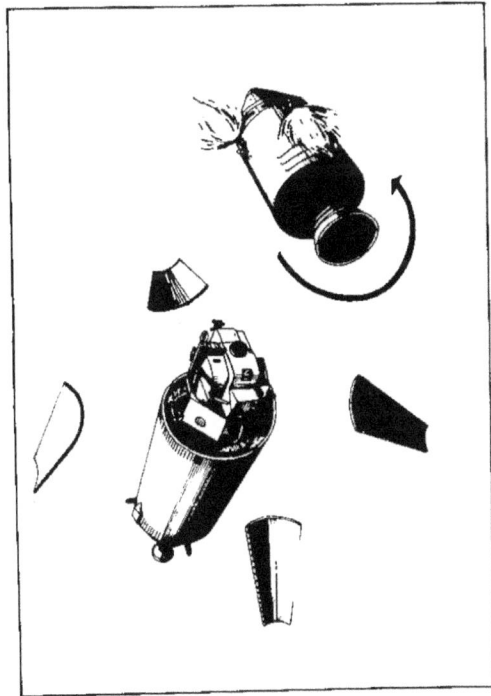

Transposition Maneuver

Navigation Check

After touchdown, the landing crew will first ready the lunar module for immediate ascent and then take a brief rest before depressurizing the cabin for two-man EVA about 10 hours after touchdown. Armstrong will step onto the lunar surface first, followed by Aldrin some 40 minutes later.

During their two hours and 40 minutes on the surface, Armstrong and Aldrin will gather geologic samples for return to Earth in sealed sample return containers and set up two scientific experiments for returning Moon data to Earth long after the mission is complete.

One experiment measures moonquakes and meteoroid impacts on the lunar surface, while the other experiment is a sophisticated reflector that will mirror laser beams back to points on Earth to aid in expanding scientific knowledge both of this planet and of the Moon.

The lunar module's descent stage will serve as a launching pad for the crew cabin as the 3,500-pound-thrust ascent engine propels the LM ascent stage back into lunar orbit for rendezvous with Collins in the command/service module--orbiting 60 miles above the Moon.

APOLLO 11 — Descent To Lunar Surface

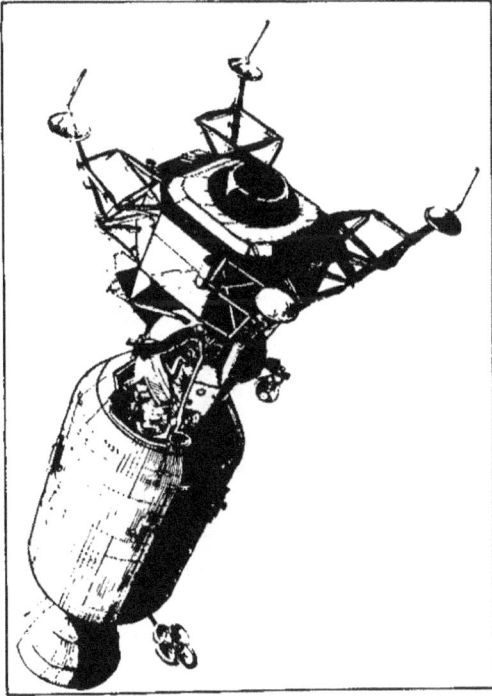

Separation Of LM From CSM

First Step On Moon

Transfer To LM

Landing On Moon

APOLLO 11 — Lunar Surface Activities

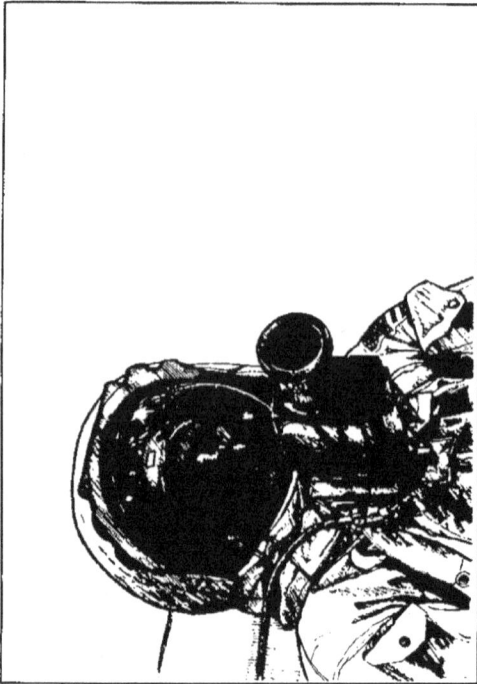

Contingency Sample

Sample Collecting

Commander On Moon

Documented Sample Collection

APOLLO 11 — Lunar Surface Activities

Bulk Sample Collection

Experiment Placements

Alignment Of Passive Seismometer

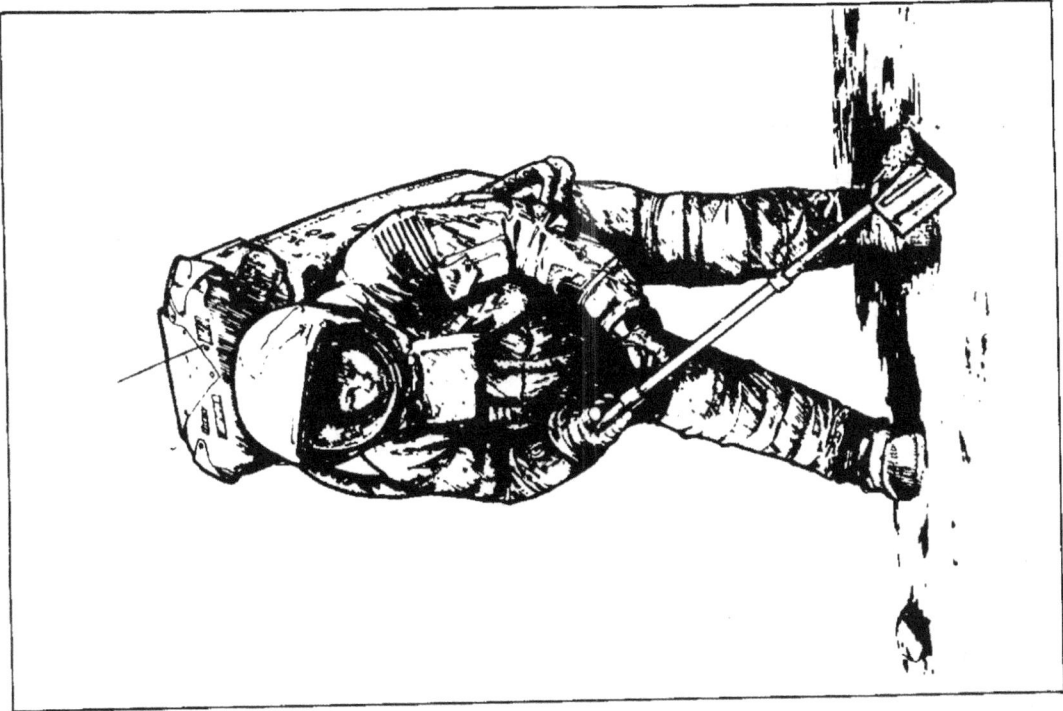

TV Camera

Four basic maneuvers, all performed by the LM crew using the spacecraft's small maneuvering and attitude thrusters, will bring the LM and the command module together for docking about three and a half hours after liftoff from the Moon.

The boost out of lunar orbit for the return journey is planned for about 135 hours after Earth liftoff and after the LM ascent stage has been jettisoned and lunar samples and film stowed aboard the command module. An optional plan provides for a 12-hour delay in the transearth injection burn to allow the crew more rest after a long hard day's work on the lunar surface and flying the rendezvous The total mission time to splashdown would remain about the same, since the transearth injection burn would impart a higher velocity to bring the spacecraft back to the mid-Pacific recovery line at about the same time.

The rendezvous sequence to be flown on Apollo 11 has twice been flown with the Apollo spacecraft---once in Earth orbit on Apollo 9 and once in lunar orbit with Apollo 10. The Apollo 10 mission duplicated, except for the actual landing, all aspects of the Apollo 11 timeline.

APOLLO 11 —— Lunar Ascent And Rendezvous

Ascent Stage Launch

LM Jettison

Return To Spacecraft

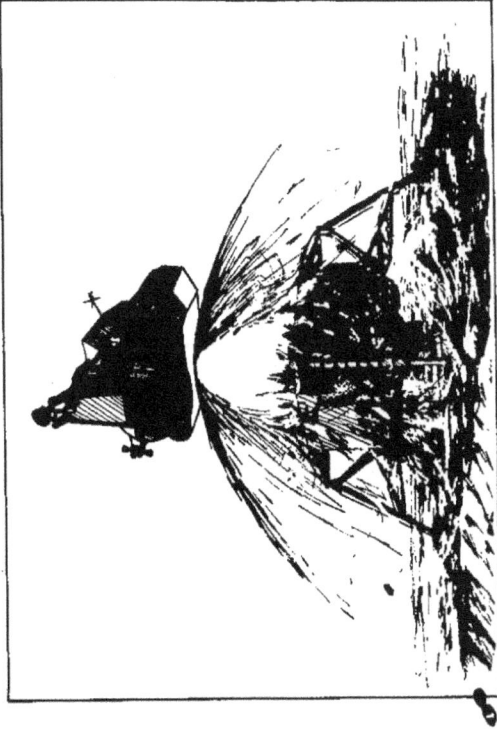

Rendezvous And Docking

APOLLO 11 — Transearth Injection And Recovery

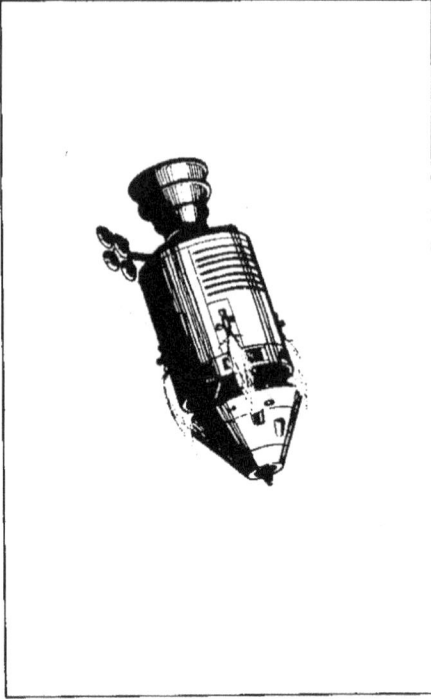
CM/SM Separation

Recovery

Transearth Injection

Reentry

Splashdown

During the transearth coast period, Apollo 11 will again control solar heat loads by using the passive thermal control "barbeque" technique. Three transearth midcourse corrections are possible and will be planned in real time to adjust the Earth entry corridor.

Apollo 11 will enter the Earth's atmosphere (400,000 feet) at 195 hours and five minutes after launch at 36,194 feet per second. Command module touchdown will be 1285 nautical miles downrange from entry at 10.6 degrees north latitude by 172.4 west longitude at 195 hours, 19 minutes after Earth launch 12:46 p.m. EDT July 24. The touchdown point is about 1040 nautical miles southwest of Honolulu, Hawaii.

(END OF GENERAL RELEASE; BACKGROUND INFORMATION FOLLOWS)

Official Apollo 11 Insignia

This photograph not for release
before Saturday, July 5, 1969

-more-

FLIGHT PROFILE

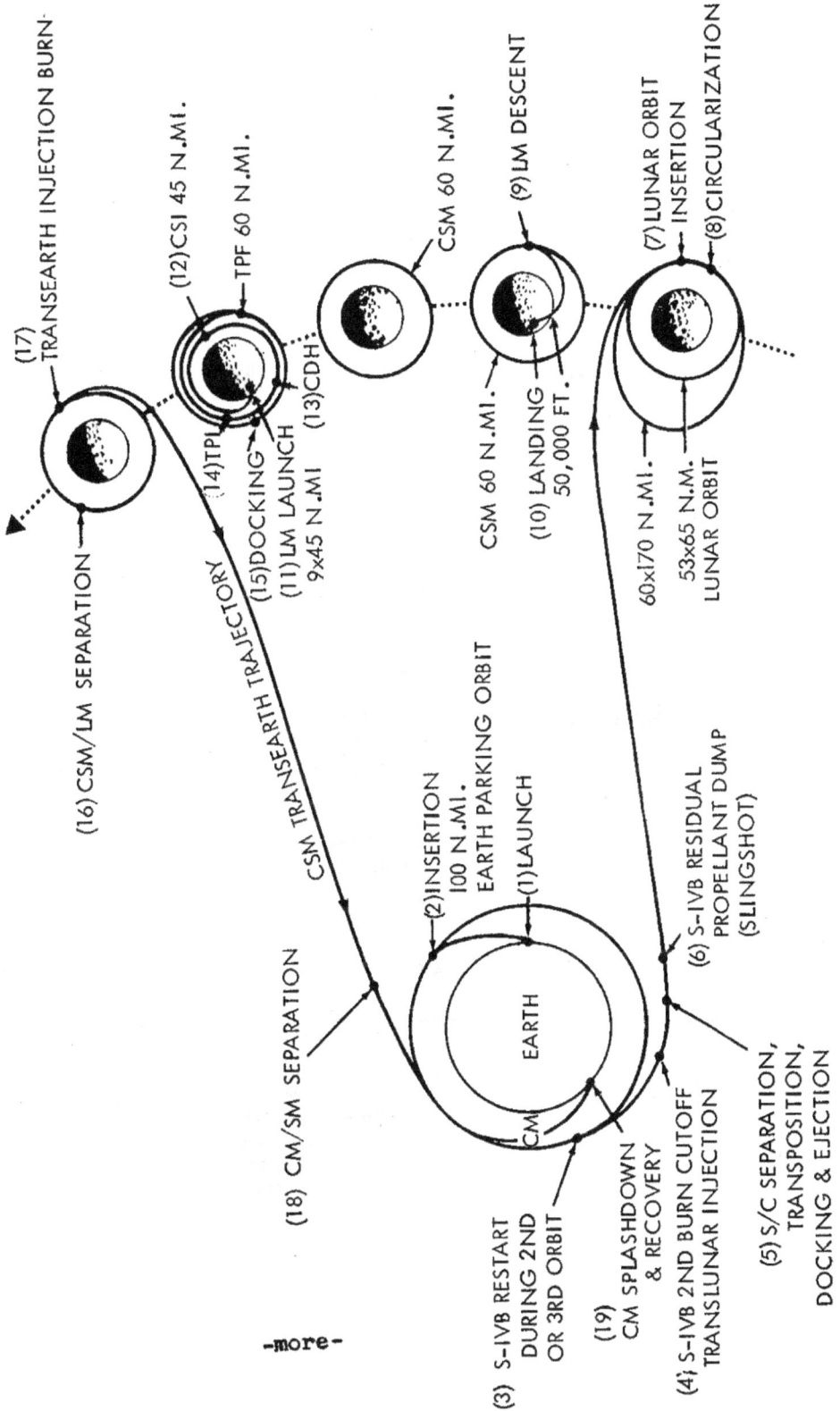

APOLLO 11 COUNTDOWN

The clock for the Apollo 11 countdown will start at T-28 hours, with a six-hour built-in-hold planned at T-9 hours, prior to launch vehicle propellant loading.

The countdown is preceded by a pre-count operation that begins some 5 days before launch. During this period the tasks include mechanical buildup of both the command/service module and LM, fuel cell activation and servicing and loading of the super critical helium aboard the LM descent stage.

Following are some of the highlights of the final count:

T-28 hrs.	Official countdown starts
T-27 hrs. 30 mins.	Install launch vehicle flight batteries (to 23 hrs. 30 mins.) LM stowage and cabin closeout (to 15 hrs.)
T-21 hrs.	Top off LM super critical helium (to 19 hrs.)
T-16 hrs.	Launch vehicle range safety checks (to 15 hrs.)
T-11 hrs. 30 mins.	Install launch vehicle destruct devices (to 10 hrs. 45 mins.) Command/service module pre-ingress operations
T-10 hrs.	Start mobile service structure move to park site
T-9 hrs.	Start six hour built-in-hold
T-9 hrs. counting	Clear blast area for propellant loading
T-8 hrs. 30 mins.	Astronaut backup crew to spacecraft for prelaunch checks
T-8 hrs. 15 mins.	Launch Vehicle propellant loading, three stages (liquid oxygen in first stage) liquid oxygen and liquid hydrogen in second, third stages. Continues thru T-3 hrs. 38 mins.

T-5 hrs. 17 mins.	Flight crew alerted
T-5 hrs. 02 mins.	Medical examination
T-4 hrs. 32 mins.	Breakfast
T-3 hrs. 57 mins.	Don space suits
T-3 hrs. 07 mins.	Depart Manned Spacecraft Operations Building for LC-39 via crew transfer van
T-2 hrs. 55 mins.	Arrive at LC-39
T-2 hrs. 40 mins.	Start flight crew ingress
T-1 hr. 55 mins.	Mission Control Center-Houston/spacecraft command checks
T-1 hr. 50 mins.	Abort advisory system checks
T-1 hr. 46 mins.	Space vehicle Emergency Detection System (EDS) test
T-43 mins.	Retrack Apollo access arm to standby position (12 degrees)
T-42 mins.	Arm launch escape system
T-40 mins.	Final launch vehicle range safety checks (to 35 mins.)
T-30 mins.	Launch vehicle power transfer test
	LM switch over to internal power
T-20 mins. to T-10 mins.	Shutdown LM operational instrumentation
T-15 mins.	Spacecraft to internal power
T-6 mins.	Space vehicle final status checks
T-5 mins. 30 sec.	Arm destruct system
T-5 mins.	Apollo access arm fully retracted
T-3 mins. 10 sec.	Initiate firing command (automatic sequencer)
T-50 sec.	Launch vehicle transfer to internal power

-more-

T-8.9 sec.   Ignition sequence start

T-2 sec.   All engines running

T-0   Liftoff

*Note: Some changes in the above countdown are possible as a
        result of experience gained in the Countdown Demonstration
        Test (CDDT) which occurs about 10 days before launch.

LAUNCH EVENTS

Time Hrs Min Sec	Event	Altitude Feet	Velocity Ft/Sec	Range Nau.Mi
00 00 00	First Motion	182.7	1,340.67	0.0
00 01 21.0	Maximum Dynamic Pressure	43,365	2,636.7	2.7
00 02 15	S-IC Center Engine Cutoff	145,600	6,504.5	24.9
00 02 40.8	S-IC Outboard Engines Cutoff	217,655	9,030.6	49.6
00 02 41.6	S-IC/S-II Separation	219,984	9,064.5	50.2
00 02 43.2	S-II Ignition	221,881	9,059.1	51.3
00 03 11.5	S-II Aft Interstage Jettison	301,266	9,469.0	87.0
00 03 17.2	LET Jettison	315,001	9,777.6	94.3
00 07 39.8	S-II Center Engine Cutoff	588,152	18,761.7	600.0
00 09 11.4	S-II Outboard Engines Cutoff	609,759	22,746.8	885.0
00 09 12.3	S-II/S-IVB Separation	609,982	22,756.7	887.99
00 09 15.4	S-IVB Ignition	610,014	22,756.7	888.42
00 11 40.1	S-IVB First Cutoff	617,957	25,562.4	1425.2
00 11 50.1	Parking Orbit Insertion	617,735	25,567.9	1463.9
02 44 14.8	S-IVB Reignition	650,558	25,554.0	3481.9
02 50 03.1	S-IVB Second Cutoff	1058,809	35,562.9	2633.6
02 50 13.1	Translunar Injection	1103,215	35,538.5	2605.0

-more-

APOLLO 11 MISSION EVENTS

Event	GET hrs:min:sec	Date/EDT	Vel.Change feet/sec	Purpose and resultant orbit
Lunar orbit insertion No. 1	75:54:28	19th 1:26 p	-2924	Inserts Apollo 11 into 60 x 170 nm elliptical lunar orbit
Lunar orbit insertion No. 2	80:09:30	19th 5:42 p	-157.8	Changes lunar parking orbit to 54 x 66 nm
CSM-LM undocking, separation (SM RCS)	100:09:50 100:39:50	20th 1:42 p 20th 2:12 p	-- 2.5	Establishes equiperiod orbit for 2.2 nm separation for DOI maneuver
Descent orbit insertion (DPS)	101:38:48	20th 3:12 p	-74.2	Lowers LM pericynthion to 8 nm (8 x 60)
LM powered descent initiation (DPS)	102:35:13	20th 4:08 p	-6761	Three-phase maneuver to brake LM out of transfer orbit, vertical descent and touchdown on lunar surface
LM touchdown on lunar surface	102:47:11	20th 4:19 p		Lunar exploration
Depressurization for lunar surface EVA	112:30	21st 2:02 a		
Repressurize LM after EVA	115:10	21st 4:42 a		

-more-

APOLLO 11 MISSION EVENTS

Event	GET hrs:min:sec	Date/EDT	Vel.Change feet/sec	Purpose and resultant orbit
Earth orbit insertion	00:11:50	16th 9:44 a	25,567	Insertion into 100 nm circular earth parking orbit
Translunar injection (S-IVB engine ignition)	02:44:15	16th 12:16 p	9,965	Injection into free-return trans-lunar trajectory with 60 nm pericynthion
CSM separation, docking	03:20:00	16th 12:52 p	--	Hard-mating of CSM and LM
Ejection from SLA	04:10:00	16th 1:42 p	1	Separates CSM-LM from S-IVB-SLA
SPS Evasive maneuver	04:39:37	16th 2:12 p	19.7	Provides separation prior to S-IVB propellant dump and "slingshot" maneuver
Midcourse correction #1	TLI+9 hrs	16th 9:16 p	*0	*These midcourse corrections have a nominal velocity change of 0 fps, but will be calculated in real time to correct TLI dispersions.
Midcourse correction #2	TLI+24 hrs	17th 12:16 p	0	
Midcourse correction #3	LOI-22 hrs	18th 3:26 p	0	
Midcourse correction #4	LOI-5 hrs	19th 8:26 a	0	

APOLLO 11 MISSION EVENTS

Event	GET hrs:min:sec	DATE/EDT	Vel.Change feet/sec	Purpose and resultant orbit
LM ascent and orbit insertion	124:23:21	21st 1:55 p	6055	Boosts ascent stage into 9 x 45 lunar orbit for rendezvous with CSM
LM RCS concentric sequence initiate (CSI) burn	125:21:20	21st 2:53 p	49.4	Raises LM perilune to 44.7 nm, adjusts orbital shape for rendezvous sequence (45.5 x 44.2)
LM RCS constant delta height (CDH) burn	126:19:40	21st 3:52 p	4.5	Radially downward burn adjusts LM orbit to constant 15 nm below CSM
LM RCS terminal phase initiate (TPI) burn	126:58:26	21st 4:30 p	24.6	LM thrusts along line of sight toward CSM, midcourse and braking maneuvers as necessary
Rendezvous (TPF)	127:43:54	21st 5:15 p	-4.7	Completes rendezvous sequence (59.5 x 59.0)
Docking	128:00:00	21st 5:32 p	--	Commander and LM pilot transfer back to CSM
LM jettison, separation (SM RCS)	131:53:05	21st 9:25 p	-1	Prevents recontact of CSM with LM ascent stage during remainder of lunar orbit
Transearth injection (TEI) SPS	135:24:34	22nd 00:57 a	3293	Inject CSM into 59.6-hour transearth trajectory

APOLLO 11 MISSION EVENTS

Event	GET hrs:min:sec	DATE/EDT	Vel.Change feet/sec	Purpose and resultant orbit
Midcourse correction No. 5	TEI+15 hrs	22nd 3:57 p	0	Transearth midcourse corrections will be computed in real time for entry corridor control and recovery area weather avoidance.
Midcourse correction No. 6	EI -15 hrs	23rd 9:37 p	0	
Midcourse correction No. 7	EI -3 hrs	24th 9:37 a	0	
CM/SM separation	194:50:04	24th 12:22 p	--	Command module oriented for entry
Entry interface (400,000 feet)	195:05:04	24th 12:37 p	--	Command module enters earth's sensible atmosphere at 36,194 fps
Touchdown	195:19:05	24th 12:51 p	--	Landing 1285 nm downrange from entry, 10.6 north latitude by 172.4 west longitude.

MISSION TRAJECTORY AND MANEUVER DESCRIPTION

Information presented herein is based upon a July 16 launch and is subject to change prior to the mission or in real time during the mission to meet changing conditions.

Launch

Apollo 11 will be launched from Kennedy Space Center Launch Complex 39A on a launch azimuth that can vary from 72 degrees to 106 degrees, depending upon the time of day of launch. The azimuth changes with time of day to permit a fuel-optimum injection from Earth parking orbit into a free-return circumlunar trajectory. Other factors influencing the launch windows are a daylight launch and proper Sun angles on the lunar landing sites.

The planned Apollo 11 launch date of July 16 will call for liftoff at 9:32 a.m. EDT on a launch azimuth of 72 degrees. The 7.6-million-pound thrust Saturn V first stage boosts the space vehicle to an altitude of 36.3 nm at 50.6 nm downrange and increases the vehicle's velocity to 9030.6 fps in 2 minutes 40.8 seconds of powered flight. First stage thrust builds to 9,088,419 pounds before center engine shutdown. Following out-board engine shutdown, the first stage separates and falls into the Atlantic Ocean about 340 nm downrange (30.3 degrees North latitude and 73.5 degrees West longitude) some 9 minutes after liftoff.

The 1-million-pound thrust second stage (S-II) carries the space vehicle to an altitude of 101.4 nm and a distance of 885 nm downrange. Before engine burnout, the vehicle will be moving at a speed of 22,746.8 fps. The outer J-2 engines will burn 6 minutes 29 seconds during this powered phase, but the center engine will be cut off at 4 minutes 56 seconds after S-II ignition.

At outboard engine cutoff, the S-II separates and, following a ballistic trajectory, plunges into the Atlantic Ocean about 2,300 nm downrange from the Kennedy Space Center (31 degrees North latitude and 33.6 degrees West longitude) some 20 minutes after liftoff.

The first burn of the Saturn V third stage (S-IVB) occurs immediately after S-II stage separation. It will last long enough (145 seconds) to insert the space vehicle into a circular Earth parking orbit beginning at about 4,818 nm downrange. Velocity at Earth orbital insertion will be 25,567 fps at 11 minutes 50 seconds ground elapsed time (GET). Inclination will be 32.6 degrees.

-more-

LAUNCH WINDOW SUMMARY

JULY 16-21

LAUNCH DATE	16	18	21
LAUNCH WINDOW, E.D.T.	9:32-13:54	9:38-14:02	10:09-14:39
SITE/PROFILE	2/FR	3/FR	5/HYB
SUN ELEVATION ANGLE	9.9-12.6	8.3-11.0	6.3-9.0
MISSION TIME, DAYS:HOURS	8:3	8:5	8:8
SPS RESERVES, FPS	1700	1550	1750

AUGUST 14-20

LAUNCH DATE	14	16	20
LAUNCH WINDOW, E.D.T.	7:51-12:15	8:04-12:31	10:05-14:47
SITE/PROFILE	2/HYB	3/HYB	5/HYB
SUN ELEVATION ANGLE	6.2-8.9	6.2-8.9	9.0-12.0
MISSION TIME, DAYS:HOURS	8:5	8:7	8:8
SPS RESERVES, FPS	1600	1750	1300

SEP 13-18

LAUNCH DATE	13	15	18
LAUNCH WINDOW, E.D.T.	6:17-10:45	7:04-11:39	11:31-16:14
SITE/PROFILE	2/HYB	3/HYB	5/HYB
SUN ELEVATION ANGLE	6.8-9.6	6.3-9.2	6.8-9.7
MISSION TIME, DAYS:HOURS	8:7	8:8	8:6
SPS RESERVES, FPS	1600	1500	1050

MISSION DURATIONS

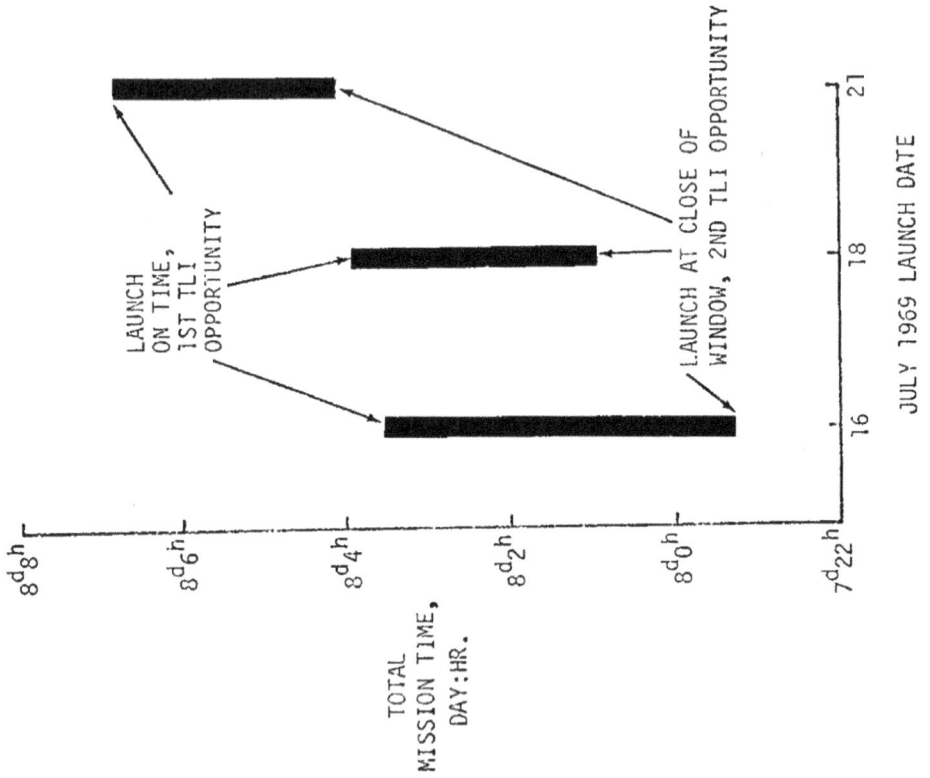

TOTAL
MISSION TIME,
DAY:HR.

LAUNCH
ON TIME,
1ST TLI
OPPORTUNITY

LAUNCH AT CLOSE OF
WINDOW, 2ND TLI OPPORTUNITY

JULY 1969 LAUNCH DATE

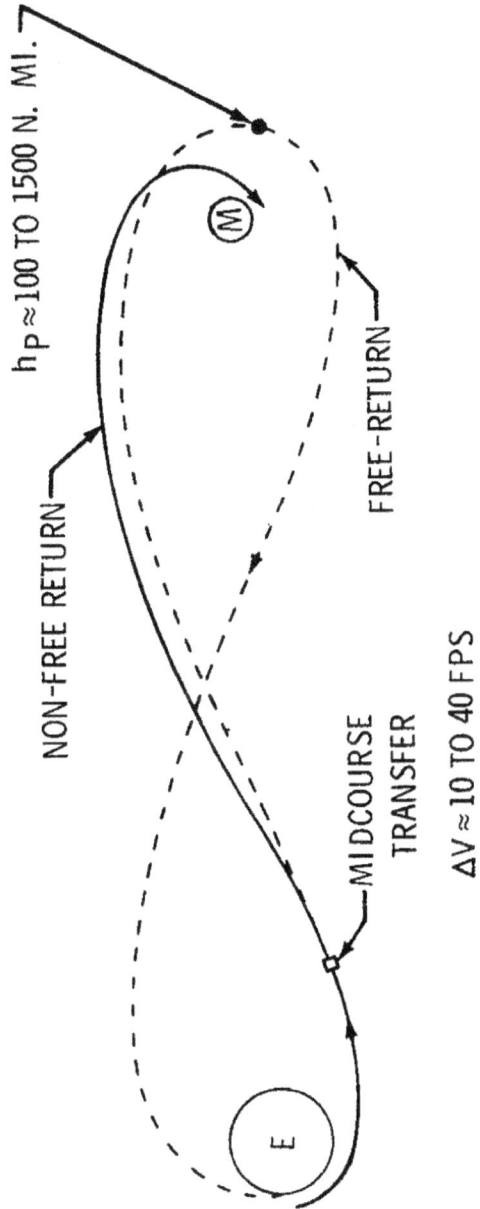

HYBRID LUNAR PROFILE

$h_p \approx 100$ TO 1500 N. MI.

NON-FREE RETURN

FREE-RETURN

MIDCOURSE TRANSFER

$\Delta V \approx 10$ TO 40 FPS

M

E

The crew will have a backup to launch vehicle guidance during powered flight. If the Saturn instrument unit inertial platform fails, the crew can switch guidance to the command module systems for first-stage powered flight automatic control. Second and third stage backup guidance is through manual takeover in which crew hand controller inputs are fed through the command module computer to the Saturn instrument unit.

Earth Parking Orbit (EPO)

Apollo 11 will remain in Earth parking orbit for one-and-one-half revolutions after insertion and will hold a local horizontal attitude during the entire period. The crew will perform spacecraft systems checks in preparation for the translunar injection (TLI) burn. The final "go" for the TLI burn will be given to the crew through the Carnarvon, Australia, Manned Space Flight Network station.

Translunar Injection (TLI)

Midway through the second revolution in Earth parking orbit, the S-IVB third-stage engine will restart at 2:44:15 GET over the mid-Pacific just south of the equator to inject Apollo 11 toward the Moon. The velocity will increase from 25,567 fps to 35,533 fps at TLI cutoff--a velocity increase of 9971 fps. The TLI burn is targeted for about 6 fps overspeed to compensate for the later SPS evasive maneuver after LM extraction. TLI will place Apollo 11 on a free-return circumlunar trajectory from which midcourse corrections if necessary could be made with the SM RCS thrusters. Entry from a free-return trajectory would be at 10:37 a.m. EDT July 22 at 14.9 degrees south latitude by 174.9 east longitude after a flight time of 145 hrs 04 min.

Transposition, Docking and Ejection (TD&E)

At about three hours after liftoff and 25 minutes after the TLI burn, the Apollo 11 crew will separate the command/service module from the spacecraft lunar module adapter (SLA), thrust out away from the S-IVB, turn around and move back in for docking with the lunar module. Docking should take place at about three hours and 21 minutes GET, and after the crew confirms all docking latches solidly engaged, they will connect the CSM-to-LM umbilicals and pressurize the LM with the command module surge tank. At about 4:09 GET, the spacecraft will be ejected from the spacecraft LM adapter by spring devices at the four LM landing gear "knee" attach points. The ejection springs will impart about one fps velocity to the spacecraft. A 19.7 fps service propulsion system (SPS) evasive maneuver in plane at 4:39 GET will separate the spacecraft to a safe distance for the S-IVB "slingshot" maneuver in which residual launch vehicle liquid propellants will be dumped through the J-2 engine bell to propel the stage into a trajectory passing behind the Moon's trailing edge and on into solar orbit.

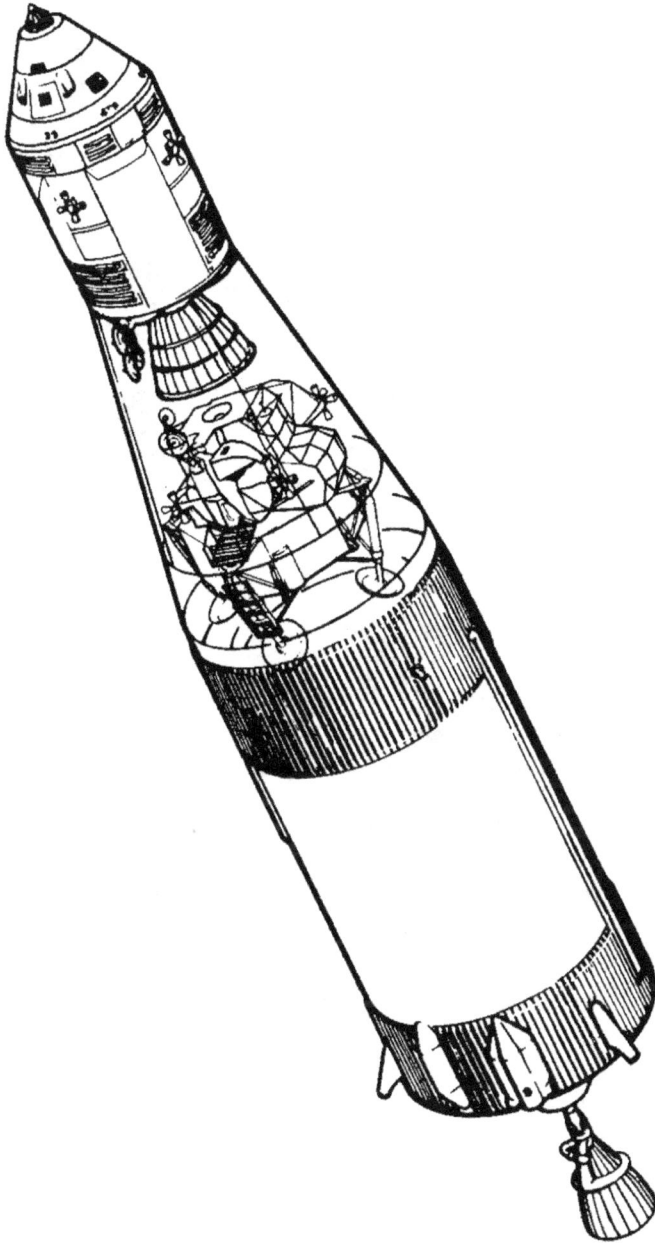

VEHICLE EARTH PARKING ORBIT CONFIGURATION
(SATURN V THIRD STAGE AND INSTRUMENT UNIT, APOLLO SPACECRAFT)

POST TLI TIMELINE

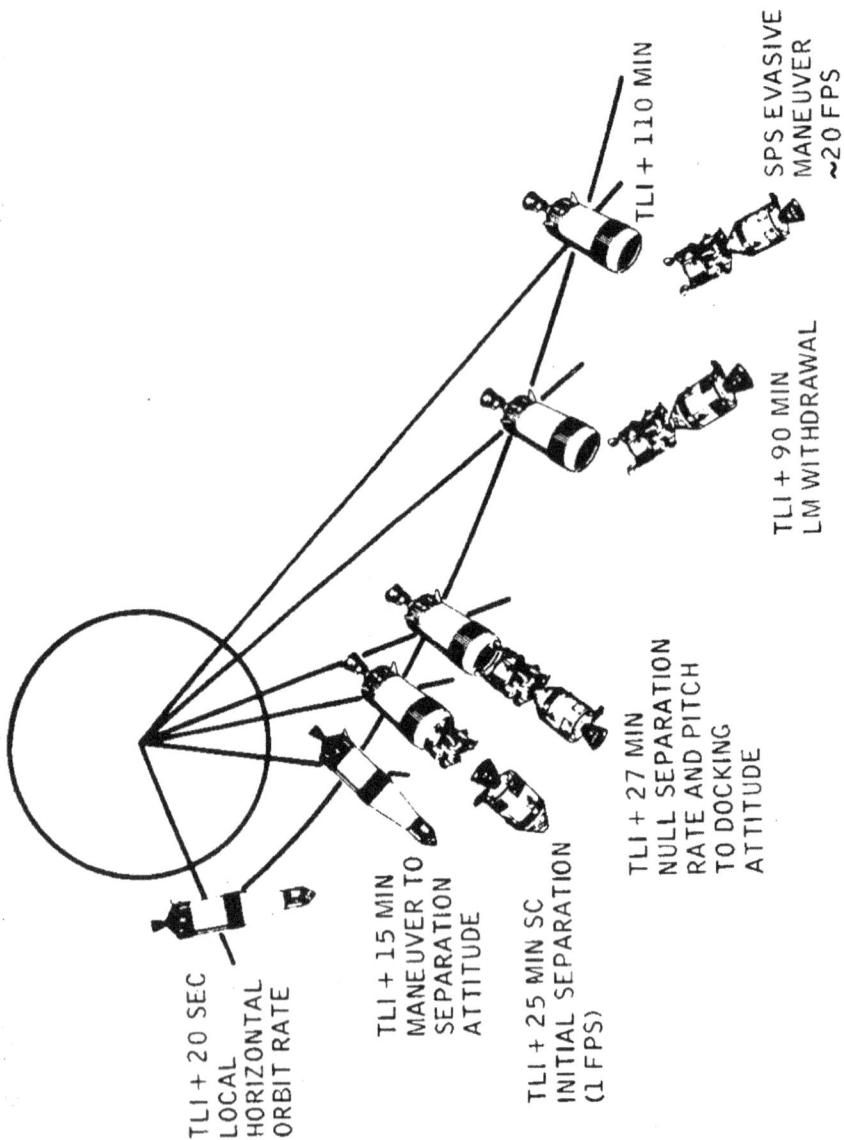

TLI + 20 SEC
LOCAL
HORIZONTAL
ORBIT RATE

TLI + 15 MIN
MANEUVER TO
SEPARATION
ATTITUDE

TLI + 25 MIN SC
INITIAL SEPARATION
(1 FPS)

TLI + 27 MIN
NULL SEPARATION
RATE AND PITCH
TO DOCKING
ATTITUDE

TLI + 90 MIN
LM WITHDRAWAL

TLI + 110 MIN

SPS EVASIVE
MANEUVER
~20 FPS

Translunar Coast

Up to four midcourse correction burns are planned during the translunar coast phase, depending upon the accuracy of the trajectory resulting from the TLI maneuver. If required, the midcourse correction burns are planned at TLI +9 hours, TLI +24 hours, lunar orbit insertion (LOI) -22 hours and LOI -5 hours.

During coast periods between midcourse corrections, the spacecraft will be in the passive thermal control (PTC) or "barbecue" mode in which the spacecraft will rotate slowly about one axis to stabilize spacecraft thermal response to the continuous solar exposure.

Lunar Orbit Insertion (LOI)

The first of two lunar orbit insertion burns will be made at 75:54:28 GET at an altitude of about 80 nm above the Moon. LOI-1 will have a nominal retrograde velocity change of 2,924 fps and will insert Apollo 11 into a 60x170-nm elliptical lunar orbit. LOI-2 two orbits later at 80:09:30 GET will adjust the orbit to a 54x65-nm orbit, which because of perturbations of the lunar gravitational potential, will become circular at 60 nm at the time of rendezvous with the LM. The burn will be 157.8 fps retrograde. Both LOI maneuvers will be with the SPS engine near pericynthion when the spacecraft is behind the Moon and out of contact with MSFN stations. After LOI-2 (circularization), the lunar module pilot will enter the lunar module for a brief checkout and return to the command module.

Lunar Module Descent, Lunar Landing

The lunar module will be manned and checked out for undocking and subsequent landing on the lunar surface at Apollo site 2. Undocking will take place at 100:09:50 GET prior to the MSFN acquisition of signal. A readially downward service module RCS burn of 2.5 fps will place the CSM on an equiperiod orbit with a maximum separation of 2.2 nm one half revolution after the separation maneuver. At this point, on lunar farside, the descent orbit insertion burn (DOI) will be made with the lunar module descent engine firing retrograde 74.2 fps at 101:38:48 GET. The burn will start at 10 per cent throttle for 15 seconds and the remainder at 40 per cent throttle.

The DOI maneuver lowers LM pericynthion to 50,000 feet at a point about 14 degrees uprange of landing site 2.

A three-phase powered descent initiation (PDI) maneuver begins at pericynthion at 102:53:13 GET using the LM descent engine to brake the vehicle out of the descent transfer orbit. The guidance-controlled PDI maneuver starts about 260 nm prior to touchdown, and is in retrograde attitude to reduce velocity to essentially zero at the time vertical descent begins. Spacecraft attitudes range from windows down at the start of PDI, to windows up as the spacecraft reaches 45,000 feet above the lunar surface and LM landing radar data can be integrated by the LM guidance computer. The braking phase ends at about 7,000 feet above the surface and the spacecraft is rotated to an upright windows-forward attitude. The start of the approach phase is called high gate, and the start of the landing phase at 500 feet is called low gate.

Both the approach phase and landing phase allow pilot take-over from guidance control as well as visual evaluation of the landing site. The final vertical descent to touchdown begins at about 150 feet when all forward velocity is nulled out. Vertical descent rate will be three fps. Touchdown will take place at 102:47:11 GET.

LUNAR ORBIT INSERTION

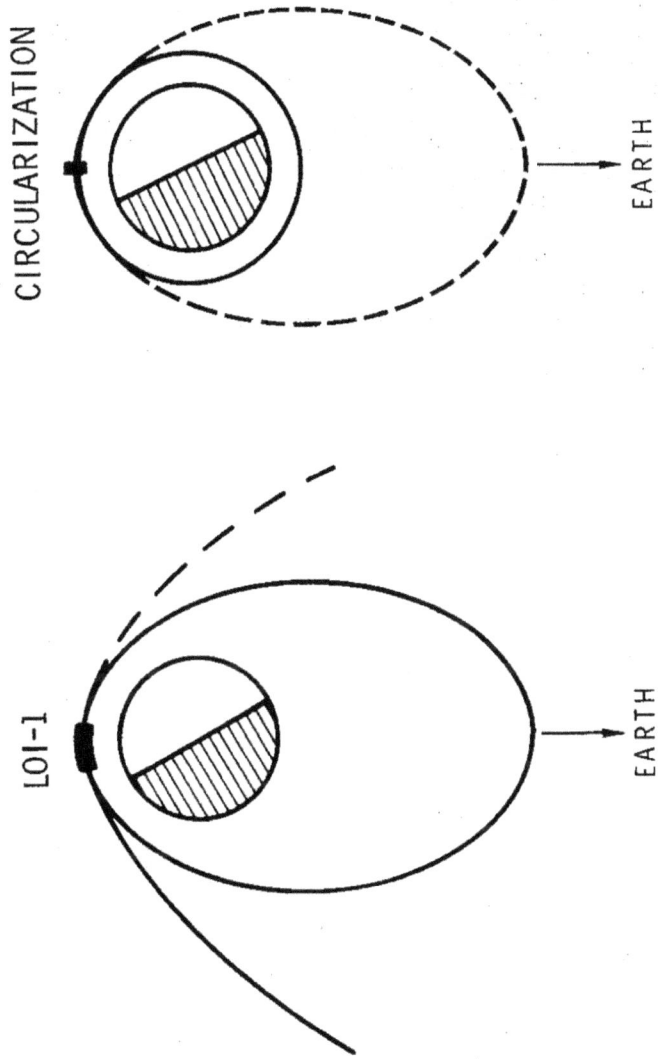

CIRCULARIZATION

LOI-1

EARTH

EARTH

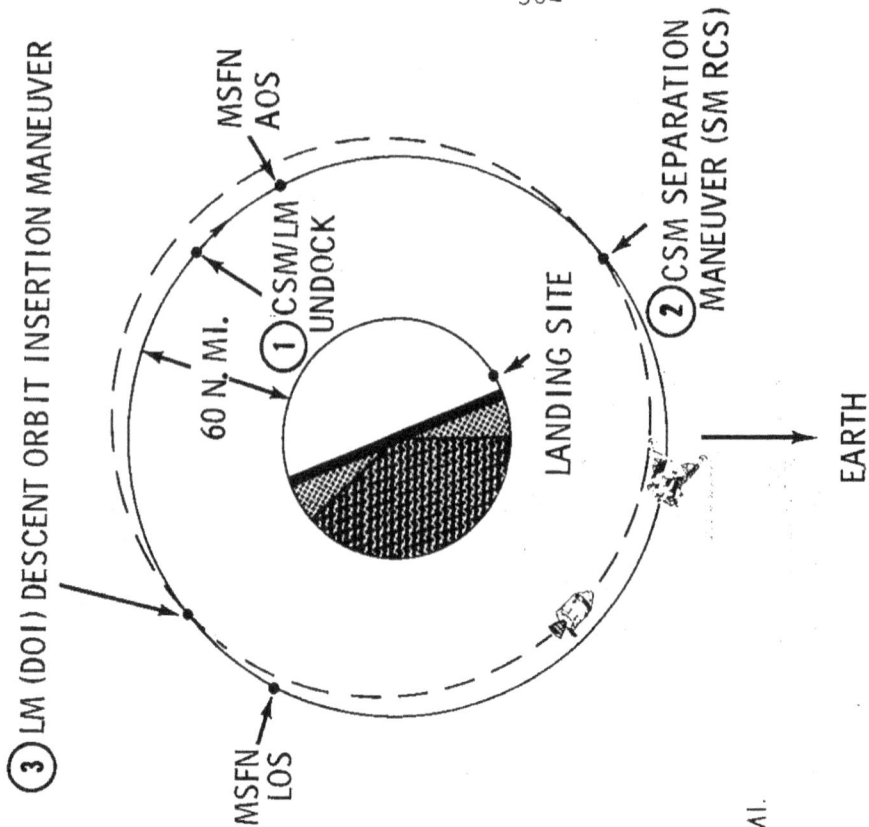

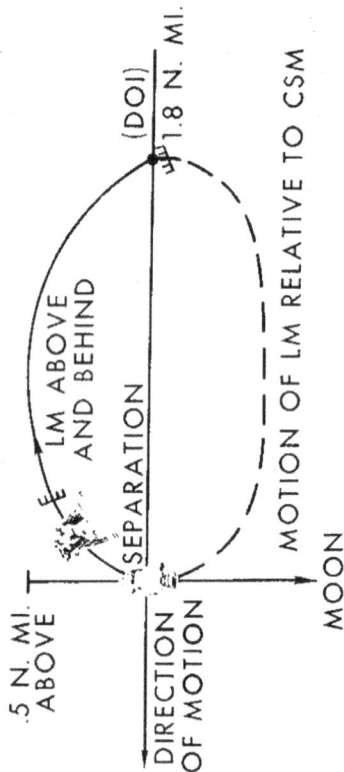

CSM/LM SEPARATION MANEUVER

LUNAR MODULE DESCENT

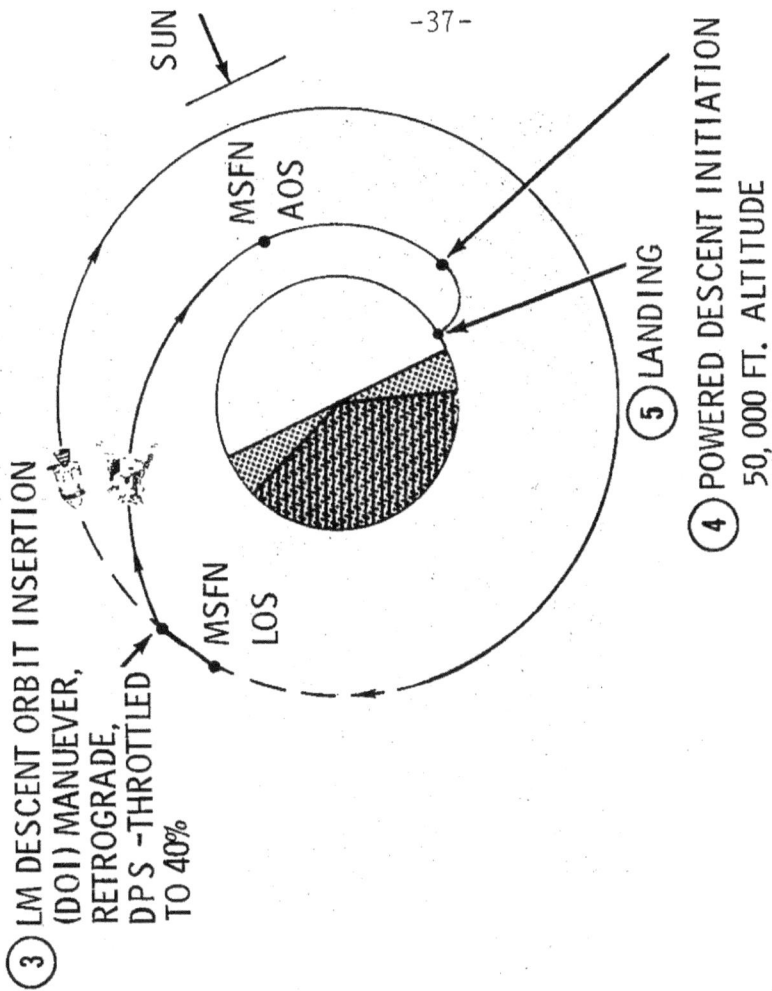

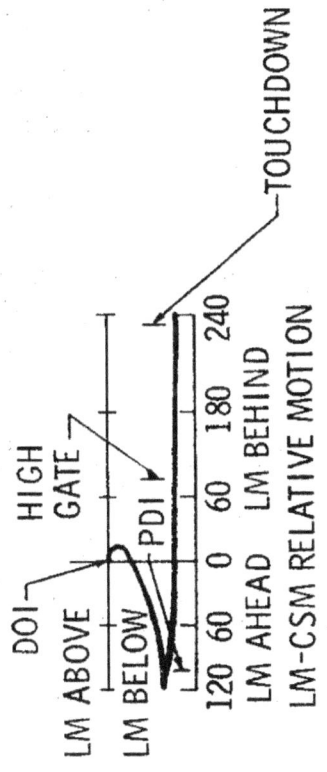

SUN

MSFN
AOS

MSFN
LOS

3 LM DESCENT ORBIT INSERTION (DOI) MANUEVER, RETROGRADE, DPS -THROTTLED TO 40%

4 POWERED DESCENT INITIATION 50,000 FT. ALTITUDE

5 LANDING

TOUCHDOWN

DOI

HIGH
GATE

PDI

LM ABOVE
LM BELOW

120 60 0 60 180 240
LM AHEAD   LM BEHIND
LM-CSM RELATIVE MOTION

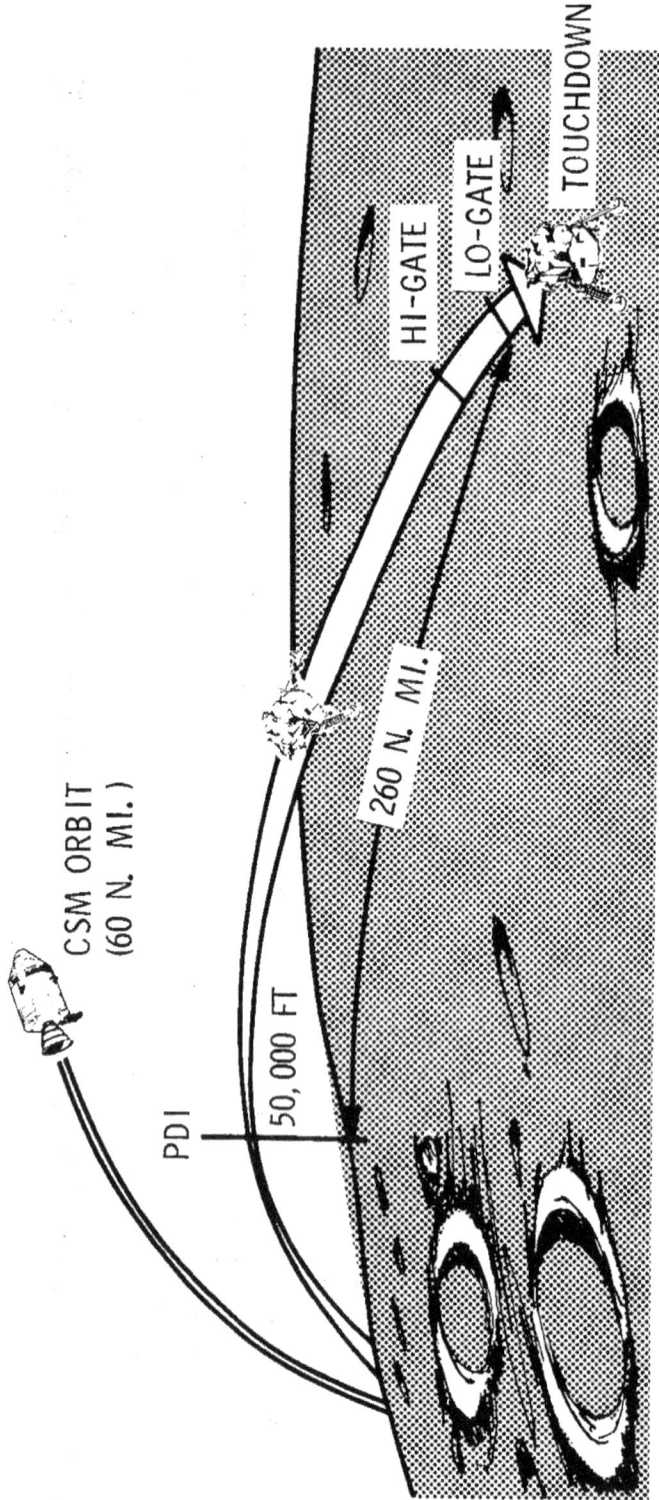

DESIGN CRITERIA

- BRAKING PHASE (PDI TO HI-GATE) - EFFICIENT REDUCTION OF ORBITAL VELOCITY
- FINAL APPROACH PHASE (HI-GATE TO LO-GATE) - CREW VISIBILITY (SAFETY OF FLIGHT AND SITE ASSESSMENT)
- LANDING PHASE (LO-GATE TO TOUCHDOWN) - MANUAL CONTROL TAKEOVER

PERATIONAL PHASES OF POWERED DESCENT

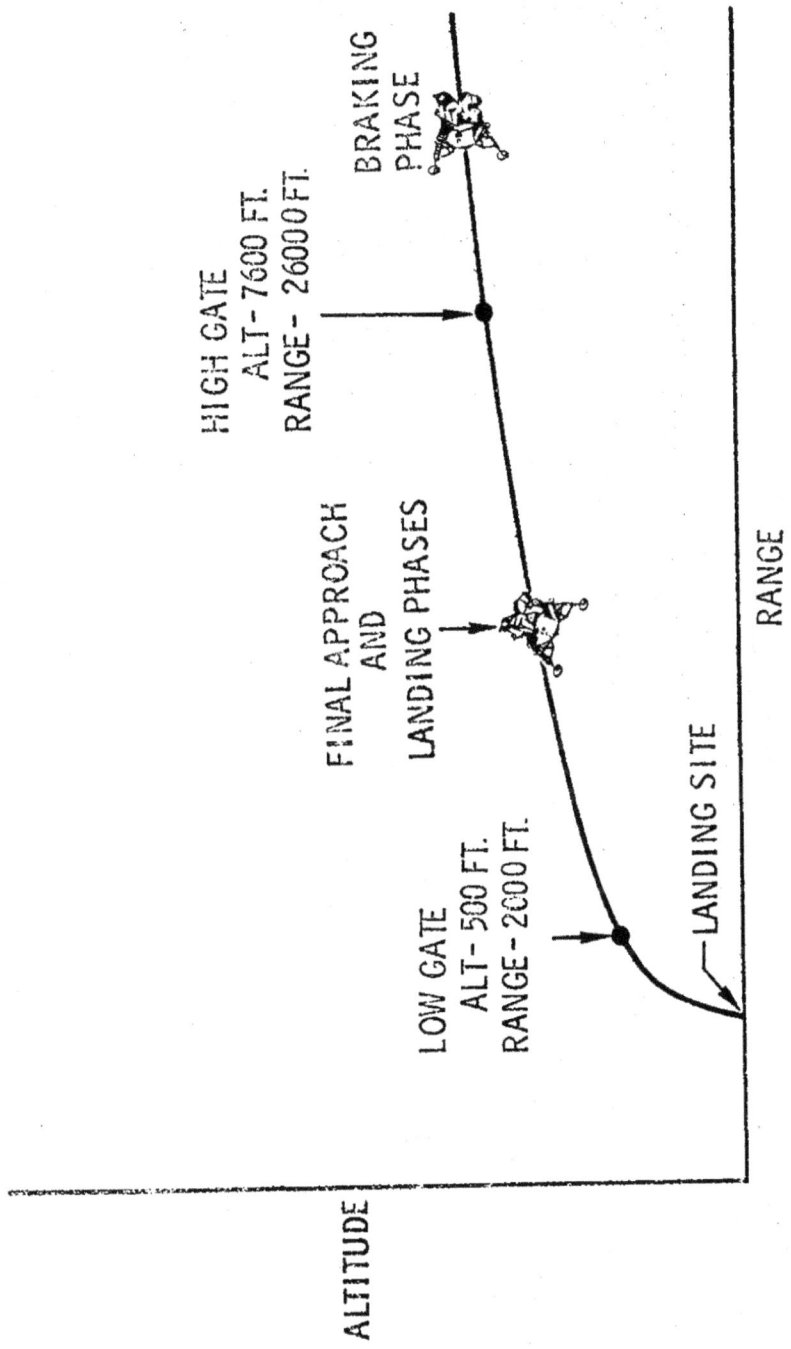

ALTITUDE

HIGH GATE
ALT - 7600 FT.
RANGE - 26000 FT.

BRAKING
PHASE

FINAL APPROACH
AND
LANDING PHASES

LOW GATE
ALT - 500 FT.
RANGE - 2000 FT.

LANDING SITE

RANGE

TARGET SEQUENCE FOR AUTOMATIC GUIDANCE

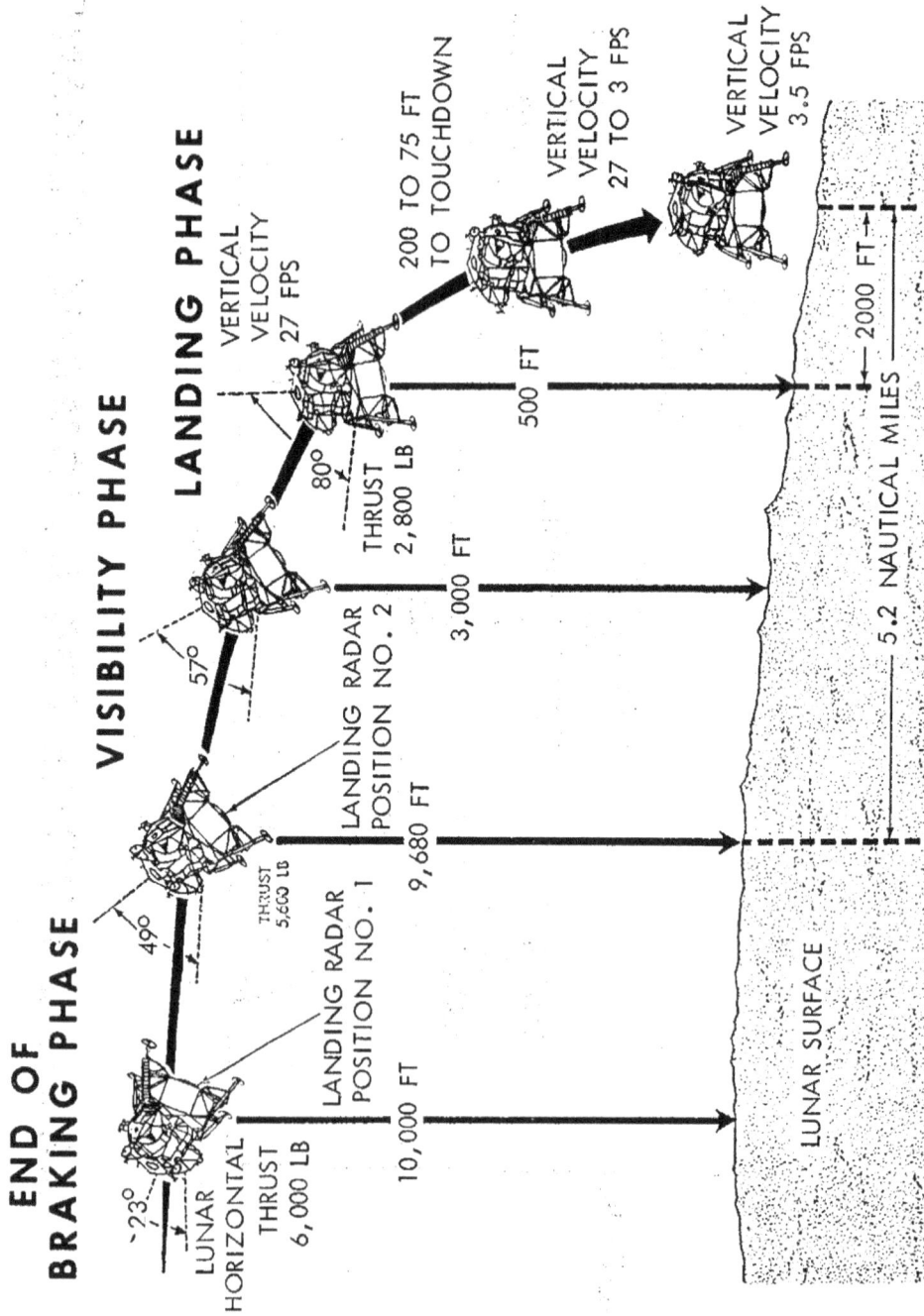

NOMINAL DESCENT TRAJECTORY FROM HIGH GATE TO TOUCHDOWN

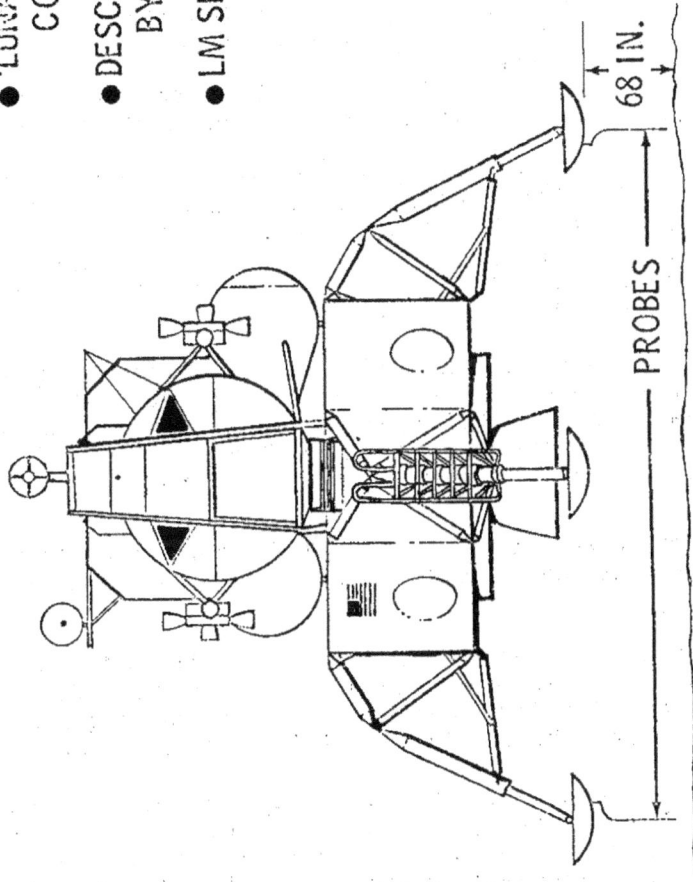

- PROBE CONTACTS LUNAR SURFACE
- 'LUNAR CONTACT' INDICATOR ON CONTROL PANEL LIGHTS
- DESCENT ENGINE IS SHUT DOWN BY CREW AFTER 1 SECOND
- LM SETTLES TO LUNAR SURFACE

68 IN.

PROBES

LUNAR CONTACT SEQUENCE

Lunar Surface Extravehicular Activity (EVA)

Armstrong and Aldrin will spend about 22 hours on the lunar surface after lunar module touchdown at 102:47:11 GET. Following extensive checkout of LM systems and preparations for contingency ascent staging, the LM crew will eat and rest before depressurizing the LM for lunar surface EVA. Both crewmen will don portable life support system (PLSS) backpacks with oxygen purge system units (OPS) attached.

LM depressurization is scheduled for 112:30 GET with the commander being the first to egress the LM and step onto the lunar surface. His movements will be recorded on still and motion picture film by the lunar module pilot and by TV deployed by the commander prior to descending the ladder. The LM pilot will leave the LM about 25 minutes after the commander and both crewmen will collect samples of lunar material and deploy the Early Apollo Scientific Experiments Package (EASEP) and the solar wind composition (SWC) experiment.

The commander, shortly after setting foot on the lunar surface, will collect a contingency sample of surface material and place it in his suit pocket. Later both crewmen will collect as much as 130 pounds of loose materials and core samples which will be stowed in air-tight sample return containers for return to Earth.

Prior to sealing the SRC, the SWC experiment, which measures the elemental and isotopic constituents of the noble (inert) gases in the solar wind, is rolled up and placed in the container for return to Earth for analysis. Principal experimenter is Dr. Johannes Geiss, University of Bern, Switzerland.

The crew will photograph the landing site terrain and inspect the LM during the EVA. They can range out to about 100 feet from the LM.

After both crewmen have ingressed the LM and have connected to the cabin suit circuit, they will doff the PLSS backpacks and jettison them along with other gear no longer needed, through the LM front hatch onto the lunar surface.

The LM cabin will be repressurized about 2 hrs. 40 min. after EVA initiation to permit transfer by the crew to the LM life support systems. The LM will then be depressurized to jettison unnecessary equipment to the lunar surface and be repressurized. The crew will have a meal and rest period before preparing for ascent into lunar orbit and rendezvousing with the CSM.

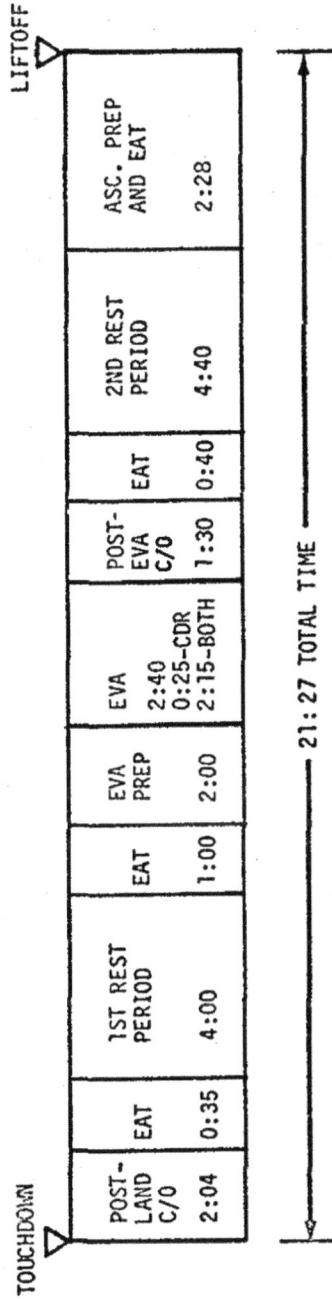

LUNAR SURFACE ACTIVITY SCHEDULE

NOMINAL EVA TIMELINE

TIME HRS + MIN

CDR

PLSS CHECKOUT	INITIAL EVA	ENVIR FAMIL	CONTG SMPL	PREL CKS	PHOTO LMP	TV DEPLOYMENT	BULK SAMPLE COLLECTION

LM PILOT

PLSS CHECKOUT	SAFETY MONITOR	MONITOR & OPERATE SEQUENCE CAMERA	INTL EVA	ENVIR FAMIL	SWC DEP	EVA & ENVIR EVALUATION

TIME: 0 10 20 30 40 50 1+00 1+10

TIME HRS + MIN

CDR

LM INSPECTION	EASEP DEPLOYMENT	DOCUMENTED SAMPLE COLLECTION	REST PHOTO LMP PREPARE AND TRANSFER SRC'S	TERMINATE EVA

LM PILOT

LM INSPECTION	EASEP DEPLOYMENT	DOCUMENTED SAMPLE COLLECTION	TERMINATE EVA	RECEIVE SRC'S

TIME: 1+10 1+20 1+30 1+40 1+50 2+00 2+10 2+20 2+30 2+40

VIEW THRU OPTICAL CENTER OF TV LENS IN DIRECTION OF "Z"-PLANE

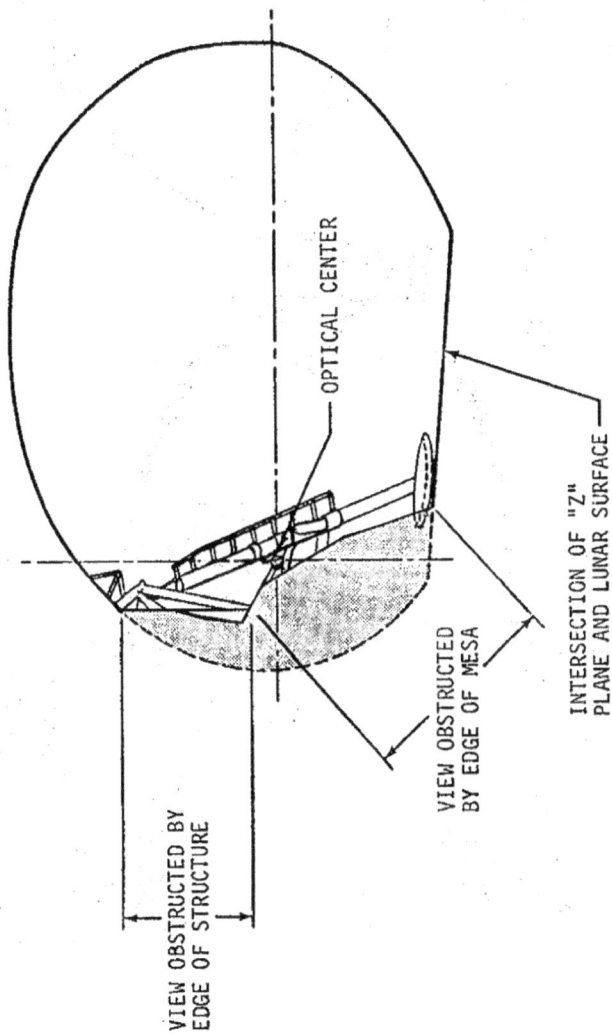

OPTICAL CENTER

INTERSECTION OF "Z"
PLANE AND LUNAR SURFACE

VIEW OBSTRUCTED
BY EDGE OF MESA

VIEW OBSTRUCTED BY
EDGE OF STRUCTURE

LUNAR SURFACE PHASE

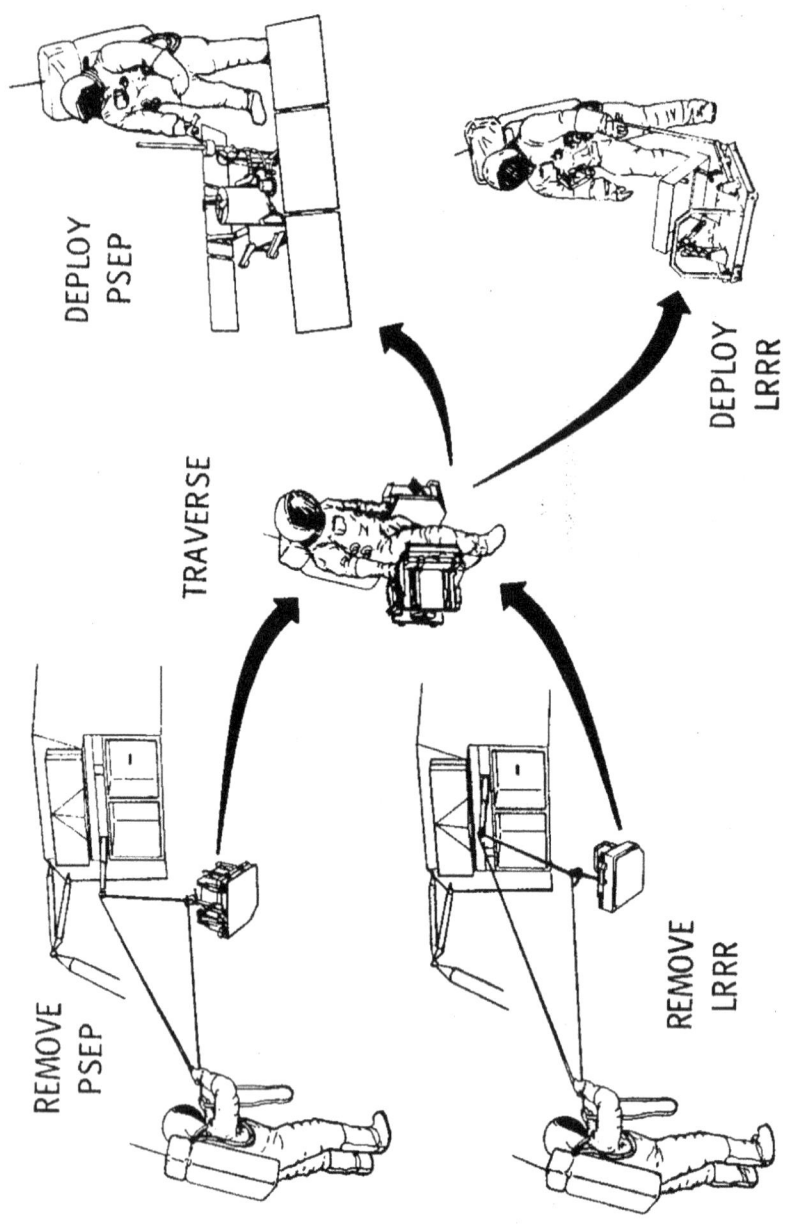

REMOVE PSEP

REMOVE LRRR

TRAVERSE

DEPLOY PSEP

DEPLOY LRRR

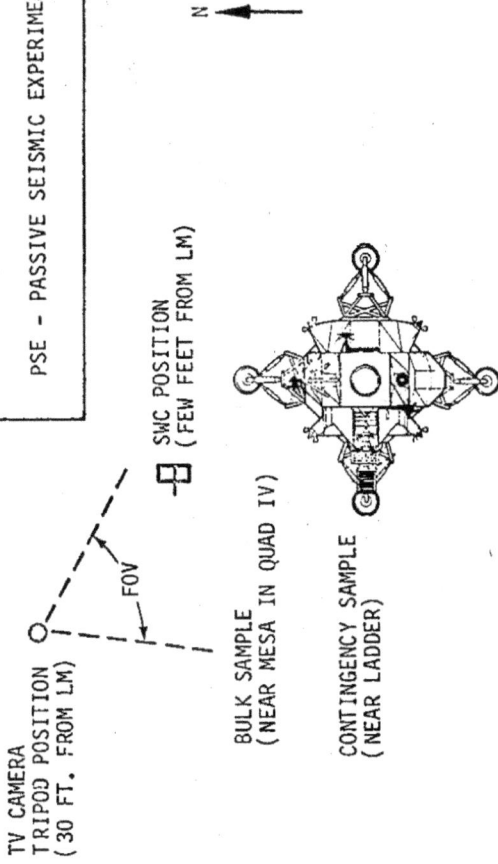

KEY:

SWC - SOLAR WIND COMPOSITION

LR3 - LASER RANGING RETRO REFLECTOR

PSE - PASSIVE SEISMIC EXPERIMENT

N

SWC POSITION
(FEW FEET FROM LM)

TV CAMERA
TRIPOD POSITION
(30 FT. FROM LM)

FOV

BULK SAMPLE
(NEAR MESA IN QUAD IV)

CONTINGENCY SAMPLE
(NEAR LADDER)

LR 3 POSITION
(70 FT. FROM LM)

PSE POSITION
(80 FT. FROM LM)

DOCUMENTED SAMPLE
(WITHIN 100 FT. FROM LM)

Lunar Sample Collection

Equipment for collecting and stowing lunar surface samples is housed in the modularized equipment stowage assembly (MESA) on the LM descent stage. The commander will unstow the equipment after adjusting to the lunar surface environment.

Items stowed in the MESA are as follows:

* Black and white TV camera.

* Large scoop for collecting bulk and documented samples of loose lunar surface material.

* Extension handle that fits the large scoop, core tubes and hammer.

* Tongs for collecting samples of rock and for picking up dropped tools.

* Gnomon for vertical reference, color and dimension scale for lunar surface photography.

* Hammer for driving core tubes, chipping rock and for trenching (with extension handle attached).

* 35mm stereo camera.

* Two sample return containers (SRC) for returning up to 130 pounds of bulk and documented lunar samples. Items such as large and small sample bags, core tubes, gas analysis and lunar environment sample containers are stowed in the SRCs. Both containers are sealed after samples have been collected, documented and stowed, and the crew will hoist them into the ascent stage by means of an equipment conveyor for transfer into the command module and subsequent return to Earth for analysis in the Lunar Receiving Laboratory.

Additionally, a contingency lunar sample return container is stowed in the LM cabin for use by the commander during the early phases of his EVA. The device is a bag attached to an extending handle in which the commander will scoop up about one liter of lunar material. He then will jettison the handle and stow the contingency sample in his pressure suit pocket.

LM Ascent, Lunar Orbit Rendezvous

Following the 22-hour lunar stay time during which the commander and lunar module pilot will deploy the Early Apollo Scientific Experiments Package (EASEP), the Solar Wind Composition (SWC) experiment, and gather lunar soil samples, the LM ascent stage will lift off the lunar surface to begin the rendezvous sequence with the orbiting CSM. Ignition of the LM ascent engine will be at 124:23:21 for a 7 min 14 sec burn with a total velocity of 6,055 fps. Powered ascent is in two phases: vertical ascent for terrain clearance and the orbital insertion phase. Pitchover along the desired launch azimuth begins as the vertical ascent rate reached 50 fps about 10 seconds after liftoff at about 250 feet in altitude. Insertion into a 9 x 45-nm lunar orbit will take place about 166 nm west of the landing site.

Following LM insertion into lunar orbit, the LM crew will compute onboard the four major maneuvers for rendezvous with the CSM which is about 255 nm ahead of the LM at this point. All maneuvers in the sequence will be made with the LM RCS thrusters. The premission rendezvous sequence maneuvers, times and velocities which likely will differ slightly in real time, are as follows:

LM ASCENT

CSM
(60 BY 60 N. MI.)

45 N. MI.

9 N. MI.

10 N. MI.

10°

8°

SUN

POWERED ASCENT INSERTION
(9/45 N. MI. ORBIT)

6

EARTH

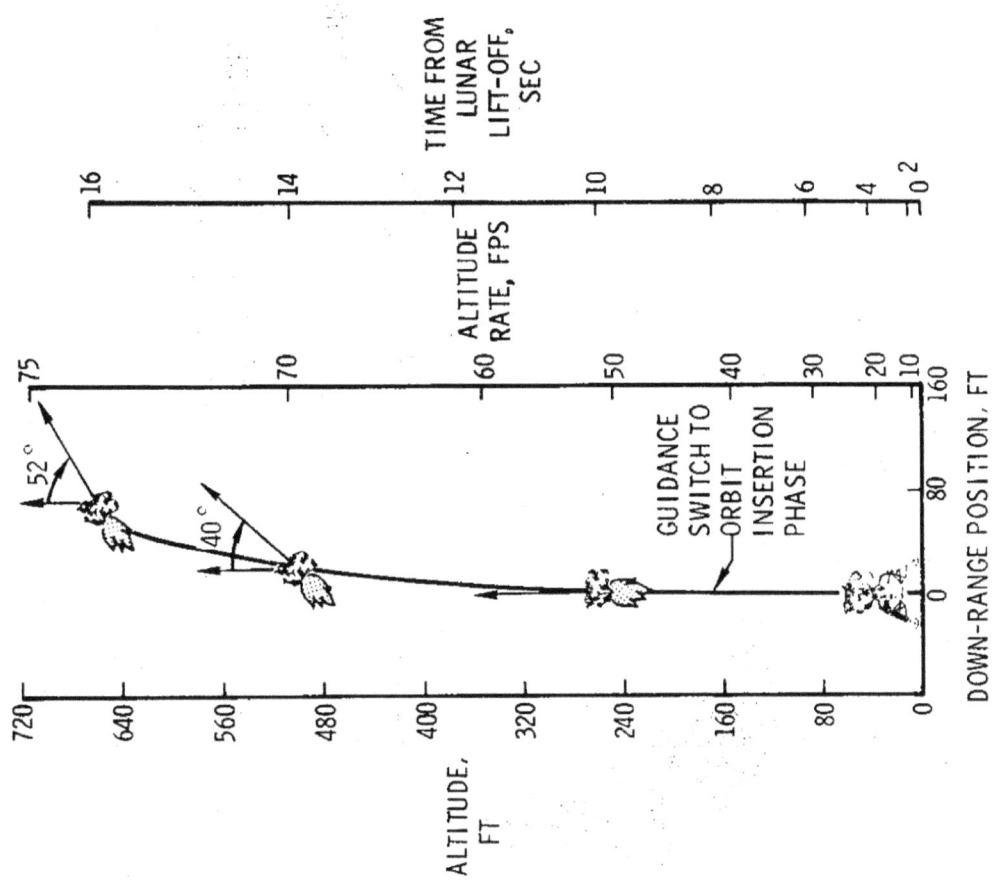

ORBIT INSERTION PHASE

END VERTICAL RISE

ORBIT INSERTION PHASE

ASCENT BURN OUT

COAST TO 44.07 N. MI. APOLUNE

59,927.5 FT

~166.52 N. MI.

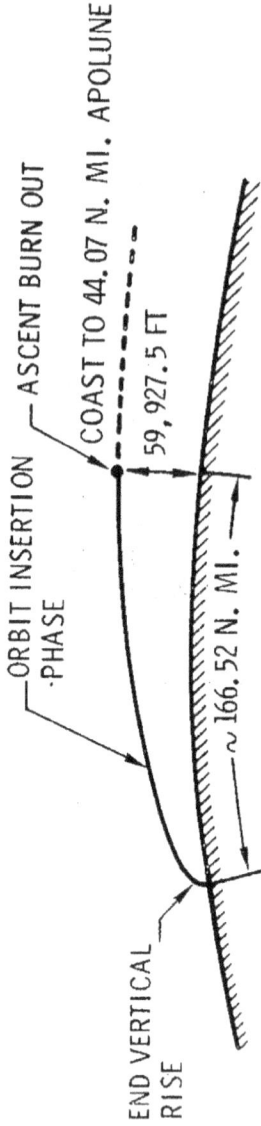

TOTAL ASCENT:
BURN TIME = 7:14.65 MIN:SEC
ΔV REQUIRED = 6,055.39 FPS
PROPELLANT REQUIRED = 4,989.86 LB

ONBOARD DISPLAYS AT INSERTION

V = 5,535.9 FPS
ḣ = 32.2 FPS
h = 60,129.5 FT

INSERTION ORBIT PARAMETERS

h_p = 55,905.4 FT

h_a = 44.07 N. MI.

η = 17.59°
γ = .324

Concentric sequence initiate (CSI): At first LM apolune after insertion 125:21:20 GET, 49 fps posigrade, following some 20 minutes of LM rendezvous radar tracking and CSM sextant/VHF ranging navigation. CSI will be targeted to place the LM in an orbit 15 nm below the CSM at the time of the later constant delta height (CDH) maneuver. The CSI burn may also initiate corrections for any out-of-plane dispersions resulting from insertion azimuth errors. Resulting LM orbit after CSI will be 45.5 x 44.2 nm and will have a catchup rate to the CSM of .072 degrees per minute.

Another plane correction is possible about 29 minutes after CSI at the nodal crossing of the CSM and LM orbits to place both vehicles at a common node at the time of the CDH maneuver at 126:19:40 GET.

Terminal phase initiate (TPI): This maneuver occurs at 126:58:26 and adds 24.6 fps along the line of sight toward the CSM when the elevation angle to the CSM reaches 26.6 degrees. The LM orbit becomes 61.2 x 43.2 nm and the catchup rate to the CSM decreases to .032 degrees per second, or a closing rate of 131 fps.

Two midcourse correction maneuvers will be made if needed, followed by four braking maneuvers at: 127:39:43 GET, 11.5 fps; 127:40:56, 9.8 fps; 127:42:35 GET, 4.8 fps; and at 127:43:54 GET, 4.7 fps. Docking nominally will take place at 128 hrs GET to end three and one-half hours of the rendezvous sequence.

Transearth Injection (TEI)

The LM ascent stage will be jettisoned about four hours after hard docking and the CSM will make a 1 fps retrograde separation maneuver.

The nominal transearth injection burn will be at 135:24 GET following 59.5 hours in lunar orbit. TEI will take place on the lunar farside and will be a 3,293 fps posigrade SPS burn of 2 min 29 sec duration and will produce an entry velocity of 36,194 fps after a 59.6 hr transearth flight time.

An optional TEI plan for five revolutions later would allow a crew rest period before making the maneuver. TEI ignition under the optional plan would take place at 145:23:45 GET with a 3,698 fps posigrade SPS burn producing an entry velocity of 36,296 fps and a transearth flight time of 51.8 hrs.

LUNAR MODULE
CONCENTRIC SEQUENCE INITIATION MANEUVER

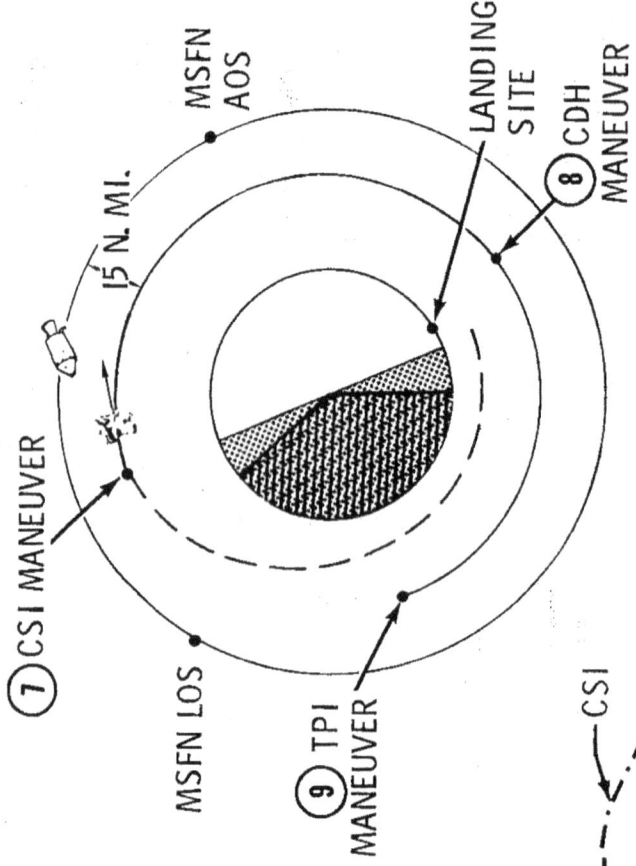

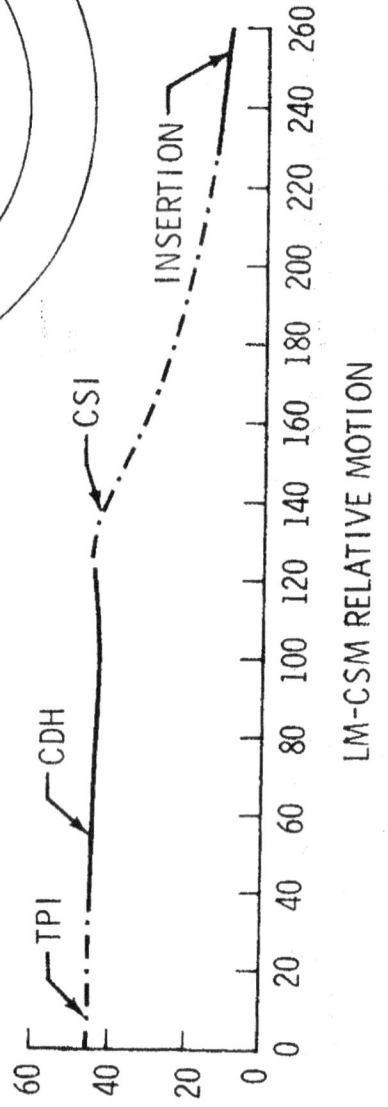

MSFN AOS

LANDING SITE

15 N. MI.

8 CDH MANEUVER

7 CSI MANEUVER

MSFN LOS

9 TPI MANEUVER

INSERTION

CSI

CDH

TPI

LM-CSM RELATIVE MOTION

LUNAR MODULE CONSTANT
DIFFERENTIAL HEIGHT AND TERMINAL PHASE MANEUVERS

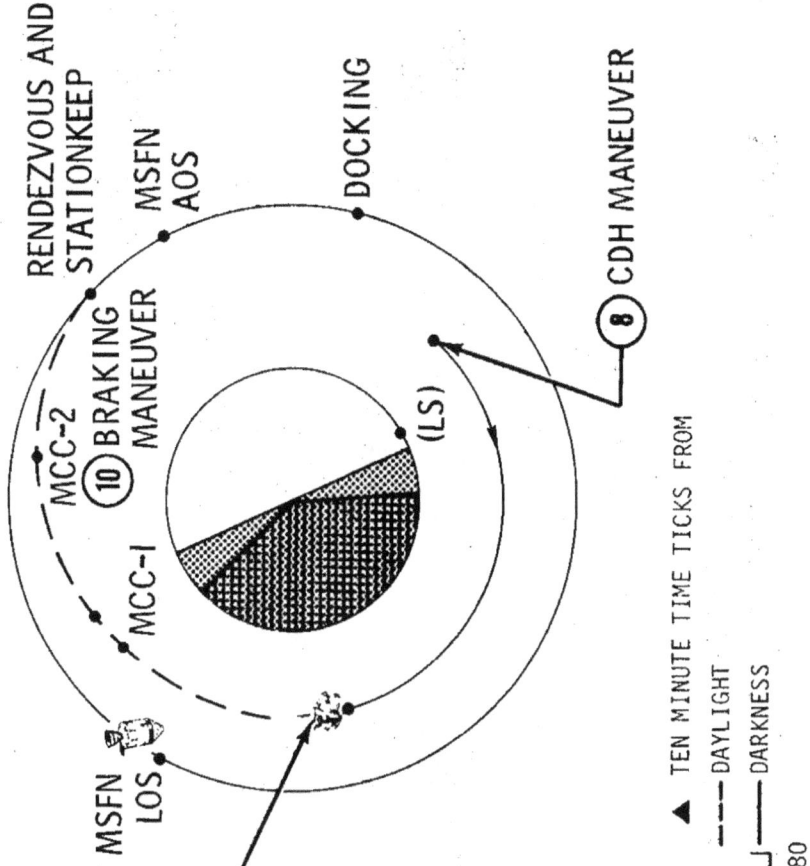

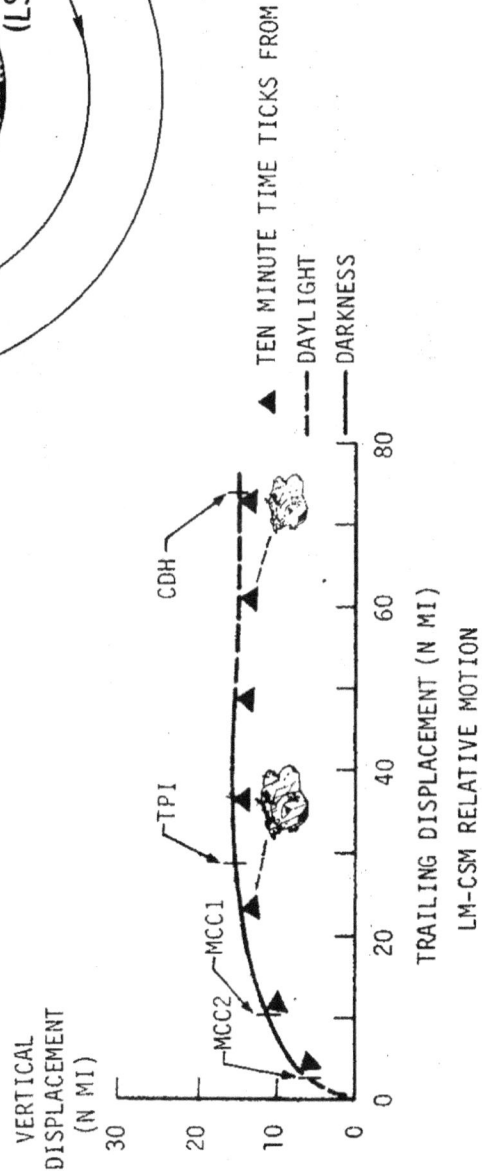

TRANSEARTH INJECTION

G. E. T. IGN   135 H 23 M

ΔV   3294 F P S

BURN TIME   2 M 29 S

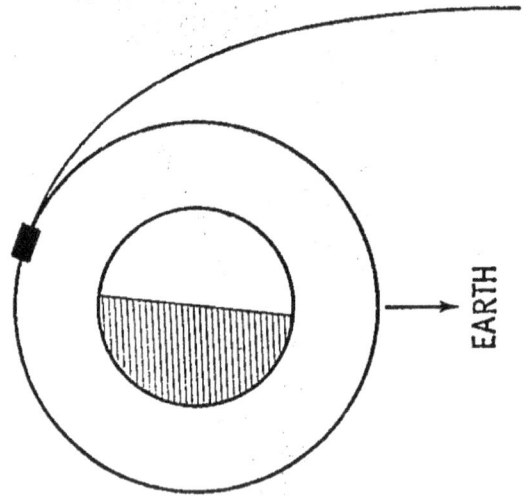

EARTH

Transearth coast

Three corridor-control transearth midcourse correction burns will be made if needed: MCC-5 at TEI +15 hrs, MCC-6 at entry interface (EI=400,000 feet) -15 hrs and MCC-7 at EI -3 hrs.

Entry, Landing

Apollo 11 will encounter the Earth's atmosphere (400,000 feet) at 195:05:04 GET at a velocity of 36,194 fps and will land some 1,285 nm downrange from the entry-interface point using the space-craft's lifting characteristics to reach the landing point. Touch-down will be at 195:19:05 at 10.6 degrees north latitude by 172.4 west longitude.

EARTH ENTRY

● ENTRY RANGE CAPABILITY - 1200 TO 2500 N. MI.

● NOMINAL ENTRY RANGE - 1285 N. MI.

● SHORT RANGE SELECTED FOR NOMINAL MISSION BECAUSE:

  ● RANGE FROM ENTRY TO LANDING CAN BE SAME FOR PRIMARY AND BACKUP CONTROL MODES

  ● PRIMARY MODE EASIER TO MONITOR WITH SHORT RANGE

● WEATHER AVOIDANCE, WITHIN ONE DAY PRIOR TO ENTRY, IS ACHIEVED USING ENTRY RANGING CAPABILITY TO 2500 N. MI.

● UP TO ONE DAY PRIOR TO ENTRY USE PROPULSION SYSTEM TO CHANGE LANDING POINT

-more-

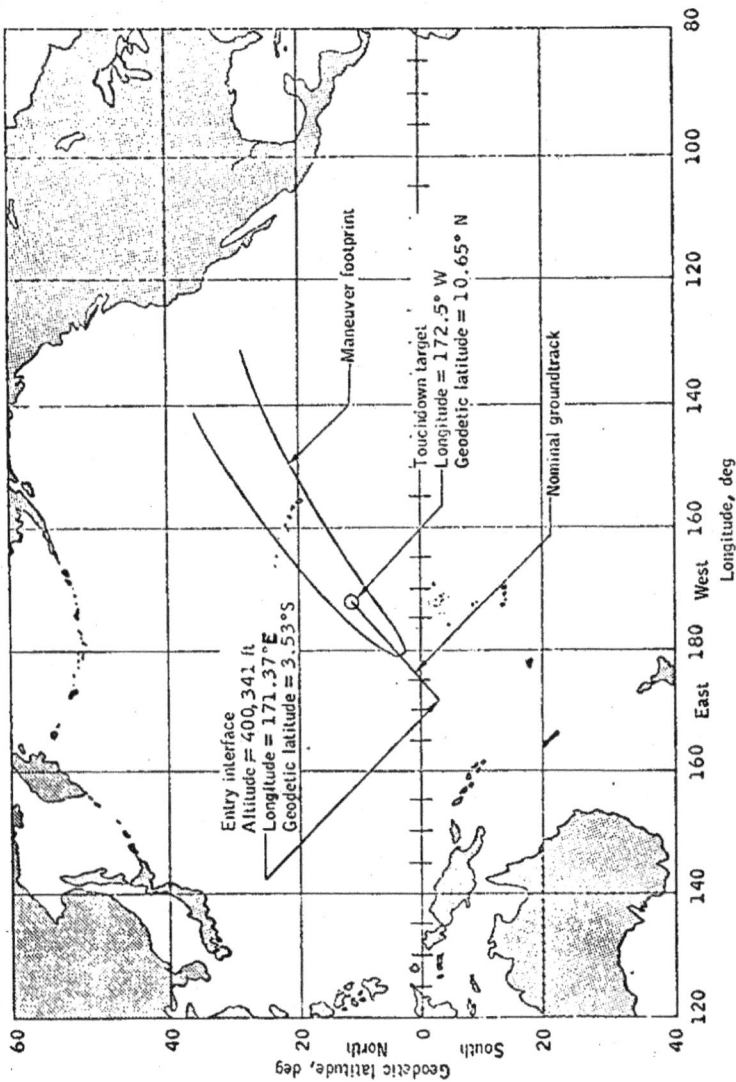

MANEUVER FOOTPRINT AND NOMINAL GROUNDTRACK

GEODETIC ALTITUDE VERSUS RANGE TO GO

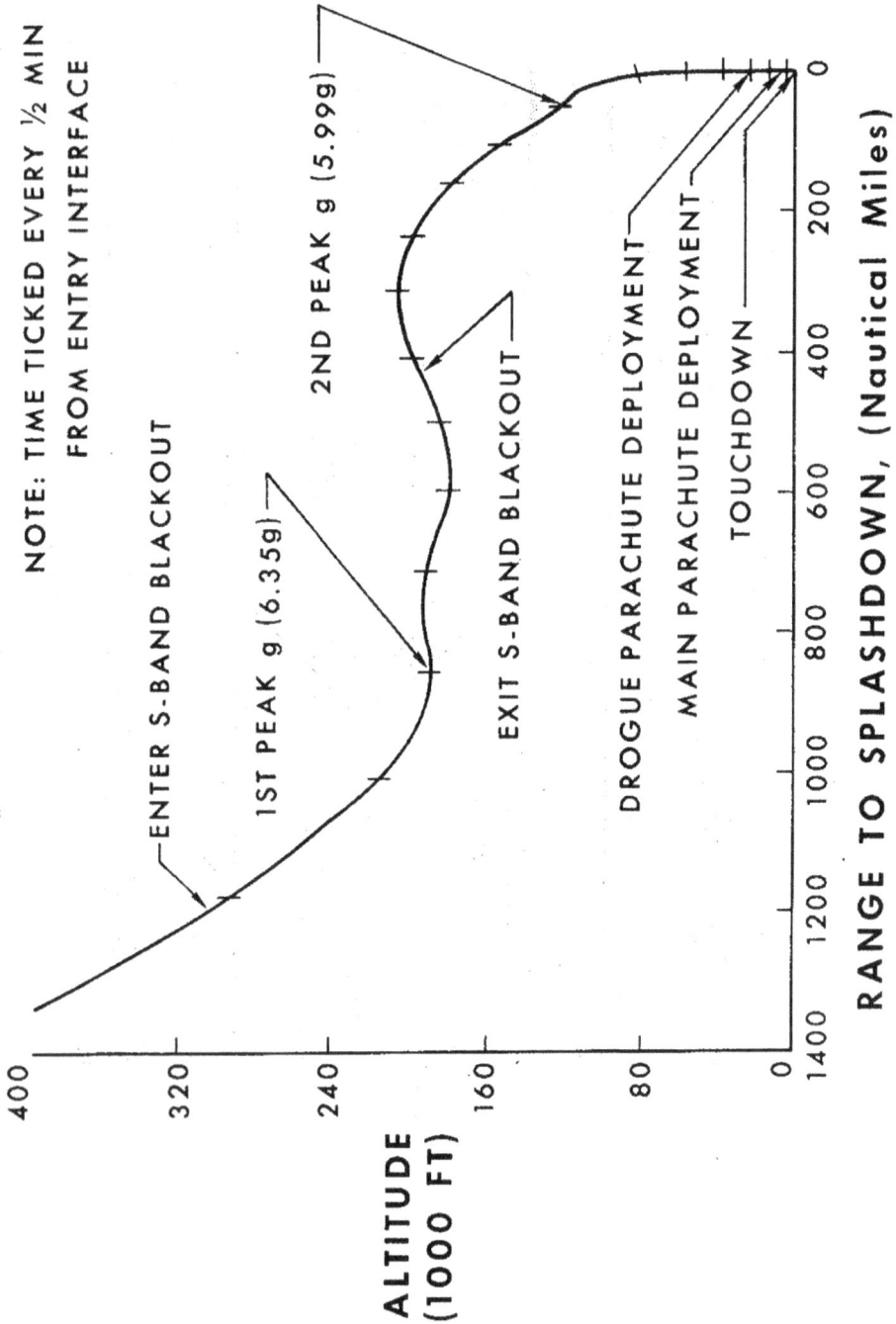

NOTE: TIME TICKED EVERY ½ MIN FROM ENTRY INTERFACE

ENTER S-BAND BLACKOUT

1ST PEAK g (6.35g)

2ND PEAK g (5.99g)

EXIT S-BAND BLACKOUT

DROGUE PARACHUTE DEPLOYMENT

MAIN PARACHUTE DEPLOYMENT

TOUCHDOWN

ALTITUDE (1000 FT)

400   320   240   160   80   0

RANGE TO SPLASHDOWN, (Nautical Miles)

1400   1200   1000   800   600   400   200   0

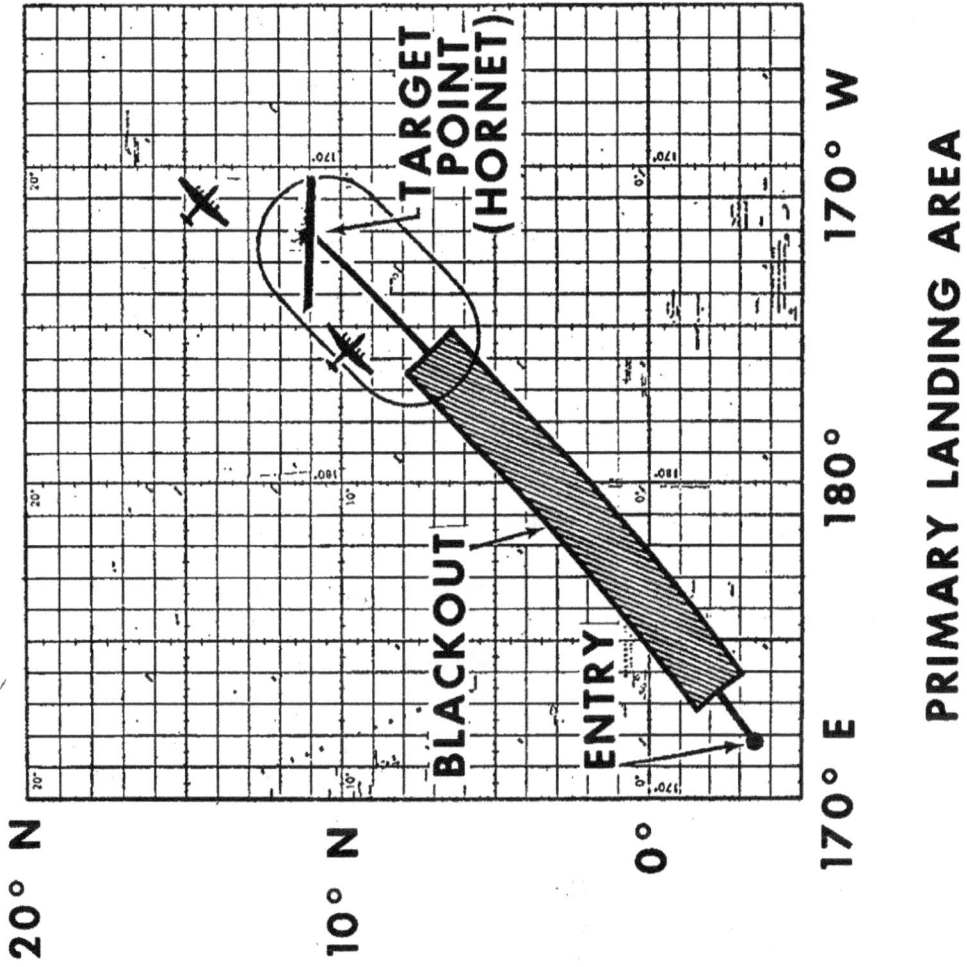

PRIMARY LANDING AREA

LIFT

DRAG

N MI (K)

2.0°

5.2°

ALT (FT)
(K)

DROGUE
CHUTES

PILOT CHUTES

DRAG
CHUTE

MAIN
CHUTES
(REEFED)

MAIN
CHUTES

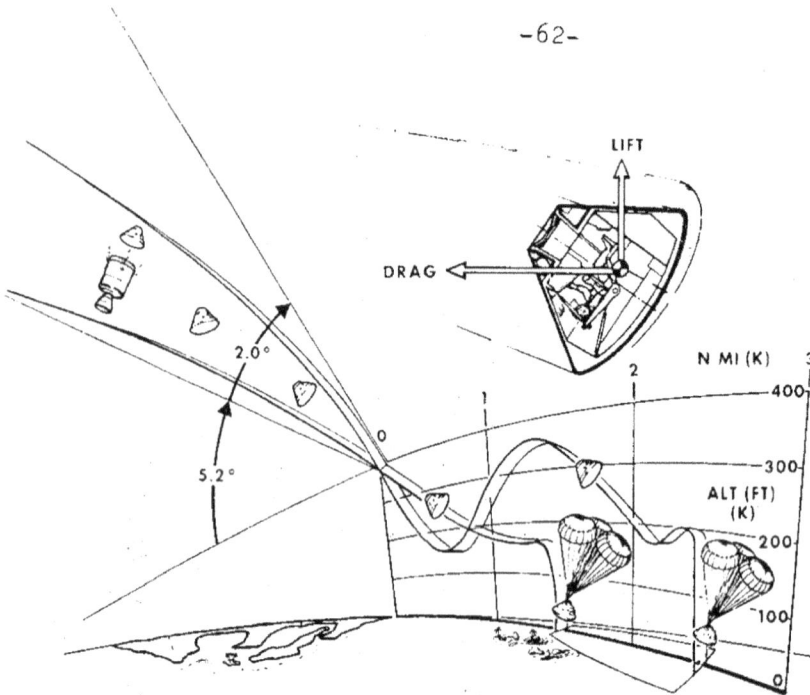

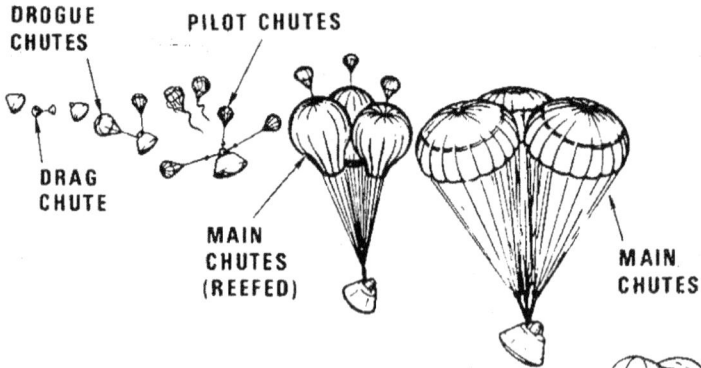

SPLASH DOWN VELOCITIES:
3 CHUTES – 31 FT/SEC
2 CHUTES – 36 FT/ SEC

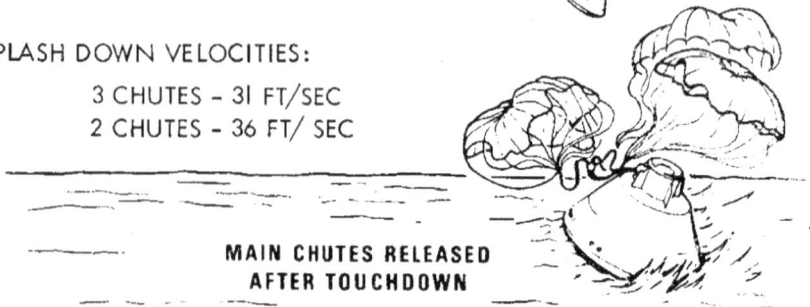

MAIN CHUTES RELEASED
AFTER TOUCHDOWN

EARTH RE-ENTRY AND LANDING

-more-

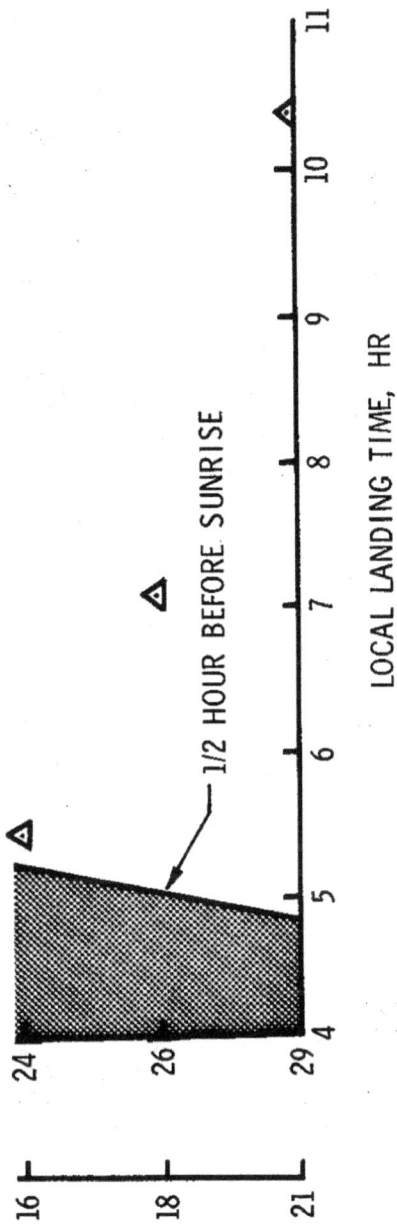

LOCAL LANDING TIMES

RECOVERY OPERATIONS, QUARANTINE

The prime recovery line for Apollo 11 is the mid-Pacific along
the 175th west meridian of longitude above 15 degrees north latitude,
and jogging to 165 degrees west longitude below the equator. The
aircraft carrier USS Hornet, Apollo 11 prime recovery ship, will be
stationed near the end-of-mission aiming point prior to entry.

Splashdown for a full-duration lunar landing mission launched
on time July 16 will be at 10.6 degrees north by 172.5 degrees west
at a ground elapsed time of 195 hrs 15 min.

The latitude of splashdown depends upon the time of the transearth injection burn and the declination of the Moon at the time of
the burn. A spacecraft returning from a lunar mission will enter the
Earth's atmosphere and splash down at a point on the Earth's farside
directly opposite the Moon. This point, called the antipode, is a
projection of a line from the center of the Moon through the center
of the Earth to the surface opposite the Moon. The mid-Pacific
recovery line rotates through the antipode once each 24 hours, and
the transearth injection burn will be targeted for splashdown along
the primary recovery line.

Other planned recovery lines for lunar missions are the East
Pacific line extending roughly parallel to the coastlines of North
and South America; the Atlantic Ocean line running along the 30th
west meridian in the northern hemisphere and along the 25th west
meridian in the southern hemisphere, and the Indian Ocean along the
65th east meridian.

Secondary landing areas for a possible Earth orbital alternate
mission are in three zones---one in the Pacific and two in the
Atlantic.

Launch abort landing areas extend downrange 3,200 nautical
miles from Kennedy Space Center, fanwise 50 nm above and below
the limits of the variable launch azimuth (72-106 degrees). Ships
on station in the launch abort area will be the destroyer USS New,
the insertion tracking ship USNS Vanguard and the minesweeper-
countermeasures ship USS Ozark.

In addition to the primary recovery ship located on the mid-
Pacific recovery line and surface vessels on the Atlantic Ocean
recovery line and in the launch abort area, 13 HC-130 aircraft
will be on standby at seven staging bases around the Earth: Guam;
Hawaii; Bermuda; Lajes, Azores; Ascension Island; Mauritius and
the Panama Canal Zone.

Apollo 11 recovery operations will be directed from the Recovery Operations Control Room in the Mission Control Center and will be supported by the Atlantic Recovery Control Center, Norfolk, Va., and the Pacific Recovery Control Center, Kunia, Hawaii.

After splashdown, the Apollo 11 crew will don biological isolation garments passed to them through the spacecraft hatch by a recovery swimmer. The crew will be carried by helicopter to the Hornet where they will enter a Mobile Quarantine Facility (MQF) about 90 minutes after landing. The MQF, with crew aboard, will be offloaded at Ford Island, Hawaii and loaded on a C-141 aircraft for the flight to Ellington AFB, Texas, and thence trucked to the Lunar Receiving Laboratory (LRL).

The crew will arrive at the LRL on July 27 following a nominal lunar landing mission and will go into the LRL Crew Reception area for a total of 21 days quarantine starting from the time they lifted off the lunar surface. The command module will arrive at the LRL two or three days later to undergo a similar quarantine. Lunar material samples will undergo a concurrent analysis in the LRL Sample Operations area during the quarantine period.

Lunar Receiving Laboratory

The Manned Spacecraft Center Lunar Receiving Laboratory has as its main function the quarantine and testing of lunar samples, spacecraft and flight crews for possible harmful organisms brought back from the lunar surface.

Detailed analysis of returned lunar samples will be done in two phases---time-critical investigations within the quarantine period and post-quarantine scientific studies of lunar samples repackaged and distributed to participating scientists.

There are 36 scientists and scientific groups selected in open world-wide competition on the scientific merits of their proposed experiments. They represent some 20 institutions in Australia, Belgium, Canada, Finland, Federal Republic of Germany, Japan, Switzerland and the United Kingdom. Major fields of investigation will be mineralogy and petrology, chemical and isotope analysis, physical properties, and biochemical and organic analysis.

The crew reception area serves as quarters for the flight crew and attendant technicians for the quarantine period in which the pilots will be debriefed and examined. The other crew reception area occupants are physicians, medical technicians, housekeepers and cooks. The CRA is also a contingency quarantine area for sample operations area people exposed to spills or vacuum system breaks.

Both the crew reception area and the sample operations area are contained within biological barrier systems that protect lunar materials from Earth contamination as well as protect the outside world from possible contamination by lunar materials.

BIOLOGICAL ISOLATION GARMENT
-more-

Analysis of lunar samples will be done in the sample operations area, and will include vacuum, magnetics, gas analysis, biological test, radiation counting and physical-chemical test laboratories.

Lunar sample return containers, or "rock boxes", will first be brought to the vacuum laboratory and opened in the ultra-clean vacuum system. After preliminary examination, the samples will be repackaged for transfer, still under vacuum, to the gas analysis, biological preparation, physical-chemical test and radiation counting laboratories.

The gas analysis lab will measure amounts and types of gases produced by lunar samples, and geochemists in the physical-chemical test lab will test the samples for their reactions to atmospheric gases and water vapor. Additionally, the physical-chemical test lab will make detailed studies of the mineralogic, petrologic, geochemical and physical properties of the samples.

Other portions of lunar samples will travel through the LRL vacuum system to the biological test lab where they will undergo tests to determine if there is life in the material that may replicate. These tests will involve introduction of lunar samples into small germ-free animals and plants. The biological test laboratory is made up of several smaller labs---bioprep, bio-analysis, germ-free, histology, normal animals (amphibia and invertebrates), incubation, anaerobic and tissue culture, crew microbiology and plants.

Some 50 feet below the LRL ground floor, the radiation counting lab will conduct low-background radioactive assay of lunar samples using gamma ray spectrometry techniques.

(See Contamination Control Program section for more details on LRL, BIGs, and the Mobile Quarantine Facility.)

SCHEDULE FOR TRANSPORT OF SAMPLES, SPACECRAFT, CREW

Samples

Two helicopters will carry lunar samples from the recovery ship to Johnston Island where they will be put aboard a C-141 and flown directly to Houston and the Lunar Receiving Laboratory (LRL). The samples should arrive at Ellington Air Force Base at about 27 hours after recovery and received in the LRL at about 9 or 10 a.m. CDT, July 25.

Spacecraft

The spacecraft is scheduled to be brought aboard the recovery ship about two hours after recovery. About 55 hours after recovery the ship is expected to arrive in Hawaii. The spacecraft will be deactivated in Hawaii (Ford Island) between 55 and 127 hours after recovery. At 130 hours it is scheduled to be loaded on a C-133 for return to Ellington AFB. Estimated time of arrival at the LRL is on July 29, 140 hours after recovery.

Crew

The flight crew is expected to enter the Mobile Quarantine Facility (MQF) on the recovery ship about 90 minutes after splashdown. The ship is expected to arrive in Hawaii at recovery plus 55 hours and the Mobile Quarantine Facility will be transferred to a C-141 aircraft at recovery plus 57 hours. The aircraft will land at Ellington AFB at recovery plus 65 hours and the MQF will arrive at the LRL about two hours later (July 27).

LUNAR RECEIVING LABORATORY PROCEDURES TIMELINE (TENTATIVE)

Sample Operations Area (SOAO)

Arrival LRL	Event	Location
Arrival	Sample containers arrive crew reception area, outer covering checked, tapes and films removed	Crew reception area
Arrival	Container #1 introduced into system	Vacuum chamber lab
	Containers weighed	" " "
	Transfer contingency sample to F-25a chamber for examination after containers #1 and #2	" " "
	Containers sterilized, dried in atmospheric decontamination and passed into glove chamber F201	" " "
	Residual gas analyzed (from containers)	" " "
" plus 5 hours	Open containers	" " "
	Weigh, preliminary exam of samples and first visual inspection by preliminary evaluation team	" " "
" plus 8 hours	Remove samples to Radiation Counting, Gas Analysis Lab &Minerology & Petrology Lab	Vacuum chamber lab RCL-Basement Min-Pet 1st floor
" plus 13 hours	Preliminary information Radiation counting. Transfer container #1 out of chamber	Vacuum chamber lab
	Initial detailed exam by Preliminary Evaluation Team Members	" " "
	Sterile sample to Bio prep (100 gms) (24 to 48 hr preparation for analysis)	Bio Test area - 1st floor
	Monopole experiment	Vacuum chamber lab

-more-

Arrival LRL	Event	Location
" plus 13 hours	Transfer samples to Phys-Chem Lab	Phys-Chem - 1st floor
	Detailed photography of samples and microscopic work	Vacuum chamber lab
" plus 24 hours	All samples canned and remain in chamber	" " "
" plus 1-2 days	Preparations of samples in bioprep lab for distribution to bio test labs. (Bacteriology, Virology, Germ-free mice) through TEI plus 21 days	Bio test labs - 1st floor
" plus 4-5 days	Early release of phys-chem analysis	Phys-Chem labs - 1st floor
" plus about 7 - 15 days	Detailed bio analysis & further phys-chem analysis	Bio test & min-pet 1st floor
" plus 15 days	Conventional samples transferred to bio test area (24-48 hours preparation for analysis)	1st floor
" plus 17 days	Bio test begins on additional bacteriological, virological, microbiological invertebrates, (fish, shrimp,oysters), birds, mice, lower invertebrates (housefly, moth, german cockroach, etc), plants (about 20) (through approximately arrival plus 30 days)	1st floor
"plus 30 days	Bio test info released on preliminary findings	1st floor
	Samples go to thin section lab (first time outside barrier) for preparation and shipment to principal investigators	1st floor

-more-

APOLLO 11 GO/NO-GO DECISION POINTS

Like Apollo 8 and 10, Apollo 11 will be flown on a step-by-step commit point or go/no-go basis in which the decisions will be made prior to each maneuver whether to continue the mission or to switch to one of the possible alternate missions. The go/no-go decisions will be made by the flight control teams in Mission Control Center jointly with the flight crew.

Go/no-go decisions will be made prior to the following events:

* Launch phase go/no-go at 10 min GET for orbit insertion

* Translunar injection

* Transposition, docking and LM extraction

* Each translunar midcourse correction burn

* Lunar orbit insertion burns Nos. 1 and 2

* CSM-LM undocking and separation

* LM descent orbit insertion

* LM powered descent initiation

* LM landing

* Periodic go/no-gos during lunar stay

* Lunar surface extravehicular activity

* LM ascent and rendezvous (A no-go would delay ascent one revolution)

* Transearth injection burn (no-go would delay TEI one or more revolutions to allow maneuver preparations to be completed)

* Each transearth midcourse correction burn.

APOLLO 11 ALTERNATE MISSIONS

Six Apollo 11 alternate missions, each aimed toward meeting the maximum number of mission objectives and gaining maximum Apollo systems experience, have been evolved for real-time choice by the mission director. The alternate missions are summarized as follows:

Alternate 1 - S-IVB fils prior to Earth orbit insertion: CSM only contingency orbit insertion (COI) with service propulsion system. The mission in Earth orbit would follow the lunar mission timeline as closely as possible and would include SPS burns similar in duration to LOI and TEI, while at the same time retaining an RCS deorbit capability. Landing would be targeted as closely as possible to the original aiming point.

Alternate 2 - S-IVB fails to restart for TLI: CSM would dock with and extract the LM as soon as possible and perform an Earth orbit mission, including docked DPS burns and possibly CSM-active rendezvous along the lunar mission timeline, with landing at the original aiming point. Failure to extract the LM would result in an Alternate 1 type mission.

Alternate 3 - No-go for nominal TLI because of orbital conditions or insufficient S-IVB propellants: TLI retargeted for lunar mission if possible; if not possible, Alternate 2 would be followed. The S-IVB would be restarted for a high-ellipse injection provided an apogee greater than 35,000 nm could be achieved. If propellants available in the S-IVB were too low to reach the 35,000 nm apogee, the TLI burn would be targeted out of plane and an Earth orbit mission along the lunar mission timeline would be flown.

Depending upon the quantity of S-IVB propellant available for a TLI-type burn that would produce an apogee greater than 35,000 nm, Alternate 3 is broken down into four subalternates:

Alternate 3A - Propellant insufficient to reach 35,000 nm

Alternate 3B - Propellant sufficient to reach apogee between 35,000 and 65,000 nm

Alternate 3C - Propellant sufficient to reach apogee between 65,000 and 200,000 nm

Alternate 3D - Propellant sufficient to reach apogee of 200,000 nm or greater; this alternate would be a near-nominal TLI burn and midcourse correction burn No. 1 would be targeted to adjust to a free-return trajectory.

Alternate 4 - Non-nominal or early shutdown TLI burn: Real-time decision would be made on whether to attempt a lunar mission or an Earth orbit mission, depending upon when TLI cutoff occurs. A lunar mission would be possible if cutoff took place during the last 40 to 45 seconds of the TLI burn. Any alternate mission chosen would include adjusting the trajectory to fit one of the above listed alternates and touchdown at the nominal mid-Pacific target point.

Alternate 5 - Failure of LM to eject after transposition and docking: CSM would continue alone for a circumlunar or lunar orbit mission, depending upon spacecraft systems status.

Alternate 6 - LM systems failure in lunar orbit: Mission would be modified in real time to gain the maximum of LM systems experience within limits of crew safety and time. If the LM descent propulsion system operated normally, the LM would be retained for DPS backup transearth injection; if the DPS were no-go, the entire LM would be jettisoned prior to TEI.

ABORT MODES

The Apollo 11 mission can be aborted at any time during the launch phase or terminated during later phases after a successful insertion into Earth orbit.

Abort modes can be summarized as follows:

Launch phase --

Mode I - Launch escape system (LES) tower propels command module away from launch vehicle. This mode is in effect from about T-45 minutes when LES is armed until LES tower jettison at 3:07 GET and command module landing point can range from the Launch Complex 39A area to 400 nm downrange.

Mode II - Begins when LES tower is jettisoned and runs until the SPS can be used to insert the CSM into a safe Earth orbit (9:22 GET) or until landing points approach the African coast. Mode II requires manual separation, entry orientation and full-lift entry with landing between 350 and 3,200 nm downrange.

Mode III - Begins when full-lift landing point reached 3,200 nm (3,560 sm, 5,931 km) and extends through Earth orbital insertion. The CSM would separate from the launch vehicle, and if necessary, an SPS retrograde burn would be made, and the command module would be flown half-lift to entry and landing at approximately 3,350 nm (3,852 sm, 6,197 km) downrange.

Mode IV and Apogee Kick - Begins after the point the SPS could be used to insert the CSM into an Earth parking orbit -- from about 9:22 GET. The SPS burn into orbit would be made two minutes after separation from the S-IVB and the mission would continue as an Earth orbit alternate. Mode IV is preferred over Mode III. A variation of Mode IV is the apogee kick in which the SPS would be ignited at first apogee to raise perigee for a safe orbit.

Deep Space Aborts

Translunar Injection Phase --

Aborts during the translunar injection phase are only a remote possibly, but if an abort became necessary during the TLI maneuver, an SPS retrograde burn could be made to produce spacecraft entry. This mode of abort would be used only in the event of an extreme emergency that affected crew safety. The spacecraft landing point would vary with launch azimuth and length of the TLI burn. Another TLI abort situation would be used if a malfunction cropped up after injection. A retrograde SPS burn at about 90 minutes after TLI shutoff would allow targeting to land on the Atlantic Ocean recovery line.

Translunar Coast phase --

Aborts arising during the three-day translunar coast
phase would be similar in nature to the 90-minute TLI abort.
Aborts from deep space bring into the play the Moon's antipode (line projected from Moon's center through Earth's Center
to the surface opposite the Moon) and the effect of the Earth's
rotation upon the geographical location of the antipode. Abort
times would be selected for landing when the 165 degree west
longitude line crosses the antipode. The mid-Pacific recovery
line crosses the antipode once each 24 hours, and if a time-
critical situation forces an abort earlier than the selected
fixed abort times, landings would be targeted for the Atlantic
Ocean, West Pacific or Indian Ocean recovery lines in that order
of preference. When the spacecraft enters the Moon's sphere
of influence, a circumlunar abort becomes faster than an attempt
to return directly to Earth.

Lunar Orbit Insertion phase --

Early SPS shutdowns during the lunar orbit insertion
burn (LOI) are covered by three modes in the Apollo 11 mission.
All three modes would result in the CM landing at the Earth
latitude of the Moon antipode at the time the abort was performed.

Mode I would be a LM DPS posigrade burn into an Earth-
return trajectory about two hours (at next pericynthion)
after an LOI shutdown during the first two minutes of the LOI
burn.

Mode II, for SPS shutdown between two and three minutes
after ignition, would use the LM DPS engine to adjust the orbit
to a safe, non-lunar impact trajectory followed by a second
DPS posigrade burn at next pericynthion targeted for the mid-
Pacific recovery line.

Mode III, from three minutes after LOI ignition until
normal cutoff, would allow the spacecraft to coast through one
or two lunar orbits before doing a DPS posigrade burn at pericynthion targeted for the mid-Pacific recovery line.

Lunar Orbit Phase --

If during lunar parking orbit it became necessary to
abort, the transearth injection (TEI) burn would be made early
and would target spacecraft landing to the mid-Pacific
recovery line.

Transearth Injection phase --

Early shutdown of the TEI burn between ignition and two minutes would cause a Mode III abort and a SPS posigrade TEI burn would be made at a later pericynthion. Cutoffs after two minutes TEI burn time would call for a Mode I abort--restart of SPS as soon as possible for Earth-return trajectory. Both modes produce mid-Pacific recovery line landings near the latitude of the antipode at the time of the TEI burn.

Transearth Coast phase --

Adjustments of the landing point are possible during the transearth coast through burns with the SPS or the service module RCS thrusters, but in general, these are covered in the discussion of transearth midcourse corrections. No abort burns will be made later than 24 hours prior to entry to avoid effects upon CM entry velocity and flight path angle.

APOLLO 11 ONBOARD TELEVISION

Two television cameras will be carried aboard Apollo 11.
A color camera of the type used on Apollo 10 will be stowed
for use aboard the command module, and the black-and-white
Apollo lunar television camera will be stowed in the LM descent stage for televising back to Earth a real-time record
of man's first step onto the Moon.

The lunar television camera weighs 7.25 pounds and draws
6.5 watts of 24-32 volts DC power. Scan rate is 10 frames-per-
second at 320 lines-per-frame. The camera body is 10.6 inches
long, 6.5 inches wide and 3.4 inches deep. The bayonet lens
mount permits lens changes by a crewman in a pressurized suit.
Two lenses, a wideangle lens for close-ups and large areas,
and a lunar day lens for viewing lunar surface features and
activities in the near field of view with sunlight illumination,
will be provided for the lunar TV camera.

The black-and-white lunar television camera is stowed in
the MESA (Modular Equipment Stowage Assembly) in the LM descent
stage and will be powered up before Armstrong starts down the
LM ladder. When he pulls the lanyard to deploy the MESA, the
TV camera will also swing down on the MESA to the left of the
ladder (as viewed from LM front) and relay a TV picture of his
initial steps on the Moon. Armstrong later will mount the TV
camera on a tripod some distance away from the LM after Aldrin
has descended to the surface. The camera will be left untended
to cover the crew's activities during the remainder of the EVA.

The Apollo lunar television camera is built by Westinghouse
Electric Corp., Aerospace Division, Baltimore, Md.

The color TV camera is a 12-pound Westinghouse camera
with a zoom lens for wideangle or close-up use, and has a three-
inch monitor which can be mounted on the camera or in the
command module. The color camera outputs a standard 525-line,
30 frame-per-second signal in color by use of a rotating color
wheel. The black-and-white signal from the spacecraft will
be converted to color at the Mission Control Center.

The following is a preliminary plan for TV passes based
upon a 9:32 a.m. EDT, July 16 launch.

TENTATIVE APOLLO 11 TV TIMES

Date	Times of Planned TV (EDT)	GET	Prime Site	Event
July 17	7:32 – 7:47 p.m.	34:00-34:15	Goldstone	Translunar Coast
July 18	7:32 – 7:47 p.m.	58:00-58:15	Goldstone	Translunar Coast
July 19	4:02 – 4:17 p.m.	78:30-78:45	Goldstone	Lunar Orbit (general surface shots)
July 20	1:52 – 2:22 p.m.	100:20-100:50	Madrid	CM/LM Formation Flying
July 21	1:57 – 2:07 a.m.	112:25-112:35	Goldstone	Landing Site Tracking
July 21	2:12 – 4:52 a.m.	112:40-115:20	*Parkes	Black and White Lunar Surface
July 22	9:02 – 9:17 p.m.	155:30-155:45	Goldstone	Transearth Coast
July 23	7:02 – 7:17 p.m.	177:30-177:45	Goldstone	Transearth Coast

* Honeysuckle will tape the Parkes pass and ship tape to MSC.

APOLLO 11 PHOTOGRAPHIC TASKS

Still and motion pictures will be made of most spacecraft maneuvers as well as of the lunar surface and of crew activities in the Apollo 11 cabin. During lunar surface activities after lunar module touchdown and the two hour 40 minute EVA, emphasis will be on photographic documentation of crew mobility, lunar surface features and lunar material sample collection.

Camera equipment carried on Apollo 11 consists of one 70mm Hasselblad electric camera stowed aboard the command module, two Hasselblad 70mm lunar surface superwide angle cameras stowed aboard the LM and a 35mm stereo close-up camera in the LM MESA.

The 2.3 pound Hasselblad superwide angle camera in the LM is fitted with a 38mm f/4.5 Zeiss Biogon lens with a focusing range from 12 inches to infinity. Shutter speeds range from time exposure and one second to 1/500 second. The angular field of view with the 38mm lens is 71 degrees vertical and horizontal on the square-format film frame.

The command module Hasselblad electric camera is normally fitted with an 80mm f/2.8 Zeiss Planar lens, but bayonet-mount 60mm and 250mm lens may be substituted for special tasks. The 80mm lens has a focusing range from three feet to infinity and has a field of view of 38 degrees vertical and horizontal.

Stowed with the Hasselblads are such associated items as a spotmeter, ringsight, polarizing filter, and film magazines. Both versions of the Hasselblad accept the same type film magazine.

For motion pictures, two Maurer 16mm data acquisition cameras (one in the CSM, one in the LM) with variable frame speed (1, 6, 12 and 24 frames per second) will be used. The cameras each weigh 2.8 pounds with a 130-foot film magazine attached. The command module 16mm camera will have lenses of 5, 18 and 75mm focal length available, while the LM camera will be fitted with the 18mm wideangle lens. Motion picture camera accessories include a right-angle mirror, a power cable and a command module boresight window bracket.

During the lunar surface extravehicular activity, the commander will be filmed by the LM pilot with the LM 16mm camera at normal or near-normal frame rates (24 and 12 fps), but when he leaves the LM to join the commander, he will switch to a one frame-per-second rate. The camera will be mounted inside the LM looking through the right-hand window. The 18mm lens has a horizontal field of view of 32 degrees and a vertical field of view of 23 degrees. At one fps, a 130-foot 16mm magazine will run out in 87 minutes in real time; projected at the standard 24 fps, the film would compress the 87 minutes to 3.6 minutes.

Armstrong and Aldrin will use the Hasselblad lunar surface camera extensively during their surface EVA to document each of their major tasks. Additionally, they will make a 360-degree overlapping panorama sequence of still photos of the lunar horizon, photograph surface features in the immediate area, make close-ups of geological samples and the area from which they were collected and record on film the appearance and condition of the lunar module after landing.

Stowed in the MESA is a 35mm stereo close-up camera which shoots 24mm square color stereo pairs with an image scale of one-half actual size. The camera is fixed focus and is equipped with a stand-off hood to position the camera at the proper focus distance. A long handle permits an EVA crewman to position the camera without stooping for surface object photography. Detail as small as 40 microns can be recorded.

A battery-powered electronic flash provides illumination. Film capacity is a minimum of 100 stereo pairs.

The stereo close-up camera will permit the Apollo 11 landing crew to photograph significant surface structure phenomena which would remain intact only in the lunar environment, such as fine powdery deposits, cracks or holes and adhesion of particles.

Near the end of EVA, the film casette will be removed and stowed in the commander's contingency sample container pocket and the camera body will be left on the lunar surface.

LUNAR DESCRIPTION

Terrain - Mountainous and crater-pitted, the former rising thousands of feet and the latter ranging from a few inches to 180 miles in diameter. The craters are thought to be formed by the impact of meteorites. The surface is covered with a layer of fine-grained material resembling silt or sand, as well as small rocks and boulders.

Environment - No air, no wind, and no moisture. The temperature ranges from 243 degrees in the two-week lunar day to 279 degrees below zero in the two-week lunar night. Gravity is one-sixth that of Earth. Micrometeoroids pelt the Moon (there is no atmosphere to burn them up). Radiation might present a problem during periods of unusual solar activity.

Dark Side - The dark or hidden side of the Moon no longer is a complete mystery. It was first photographed by a Russian craft and since then has been photographed many times, particularly by NASA's Lunar Orbiter spacecraft and Apollo 8.

Origin - There is still no agreement among scientists on the origin of the Moon. The three theories: (1) the Moon once was part of Earth and split off into its own orbit, (2) it evolved as a separate body at the same time as Earth, and (3) it formed elsewhere in space and wandered until it was captured by Earth's gravitational field.

Physical Facts

Diameter	2,160 miles (about ¼ that of Earth)
Circumference	6,790 miles (about ¼ that of Earth)
Distance from Earth	238,857 miles (mean; 221,463 minimum to 252,710 maximum)
Surface temperature	+243°F (Sun at zenith) -279°F (night)
Surface gravity	1/6 that of Earth
Mass	1/100th that of Earth
Volume	1/50th that of Earth
Lunar day and night	14 Earth days each
Mean velocity in orbit	2,287 miles per hour
Escape velocity	1.48 miles per second
Month (period of rotation around Earth)	27 days, 7 hours, 43 minutes

Apollo Lunar Landing Sites

Possible landing sites for the Apollo lunar module have been under study by NASA's Apollo Site Selection Board for more than two years. Thirty sites originally were considered. These have been narrowed down to three for the first lunar landing. (Site 1 currently not considered for first landing.)

Selection of the final sites was based on high resolution photographs by Lunar Orbiter spacecraft, plus close-up photos and surface data provided by the Surveyor spacecraft which soft-landed on the Moon.

The original sites are located on the visible side of the Moon within 45 degrees east and west of the Moon's center and 5 degrees north and south of its equator.

The final site choices were based on these factors:

*Smoothness (relatively few craters and boulders)

*Approach (no large hills, high cliffs, or deep craters that could cause incorrect altitude signals to the lunar module landing radar)

*Propellant requirements (selected sites require the least expenditure of spacecraft propellants)

*Recycle (selected sites allow effective launch preparation recycling if the Apollo Saturn V countdown is delayed)

*Free return (sites are within reach of the spacecraft launched on a free return translunar trajectory)

*Slope (there is little slope -- less than 2 degrees in the approach path and landing area)

APOLLO LUNAR LANDING SITES

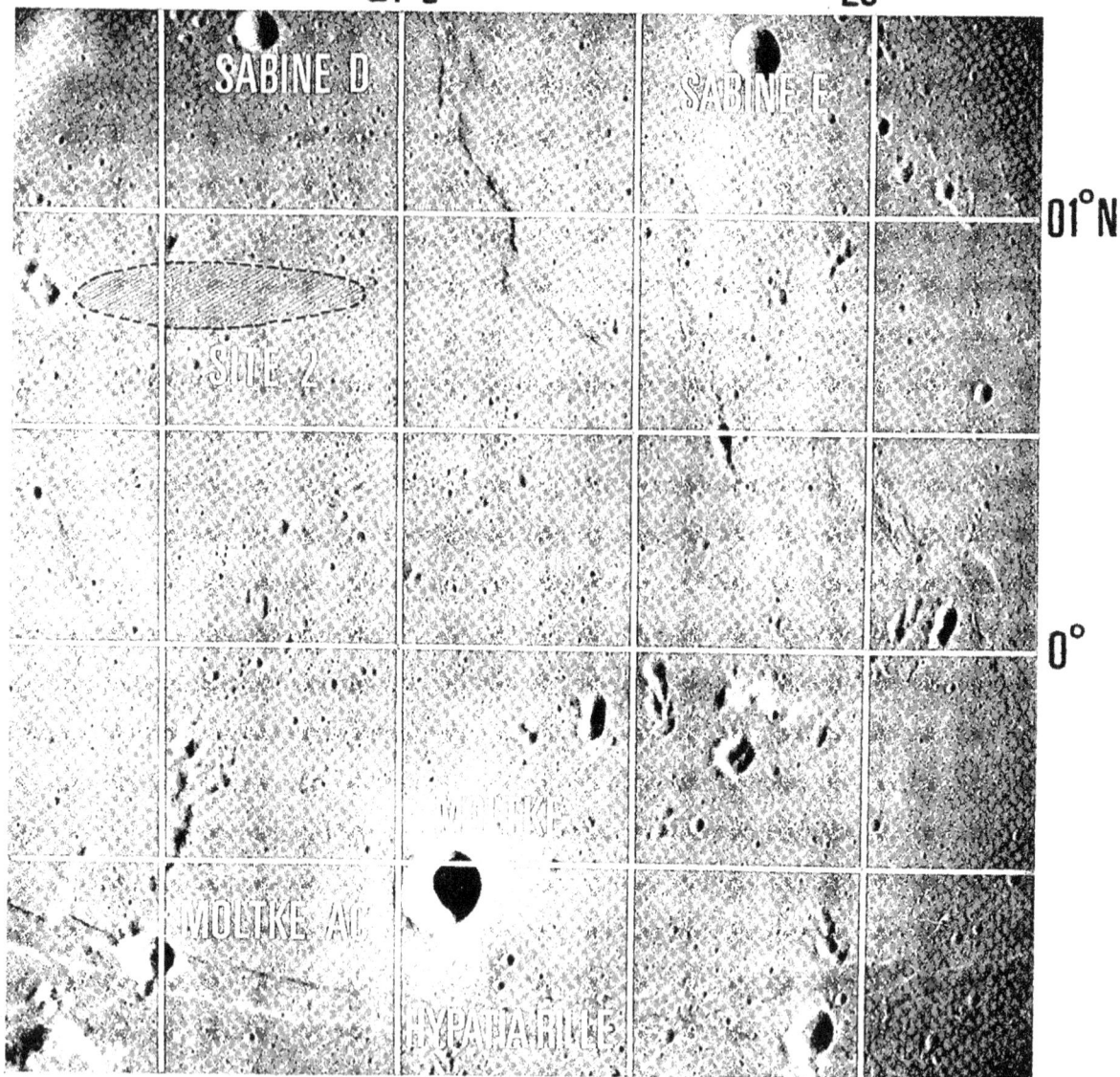

The Apollo 11 Landing Sites Are:

Site 2

latitude 0° 42' 50" North
longitude 23° 42' 28" East

Site 2 is located on the east
central part of the Moon in southwestern Mar Tranquillitatis. The
site is approximately 62 miles
(100 kilometers) east of the rim
of Crater Sabine and approximately
118 miles (190 kilometers) southwest of the Crater Maskelyne.

Site 3

latitude 0° 21' 10" North
longitude 1° 17' 57" West

Site 3 is located near the center
of the visible face of the Moon
in the southwestern part of Sinus
Medii. The site is approximately
25 miles (40 kilometers) west of
the center of the face and 21 miles
(50 kilometers) southwest of the
Crater Bruce.

Site 5

latitude 1° 40' 41" North
longitude 41° 53' 57" West

Site 5 is located on the west
central part of the visible face
in southeastern Oceanus Procel-
larum. The site is approximately
130 miles (210 kilometers) southwest of the rim of Crater Kepler
and 118 miles (190 kilometers)
north northeast of the rim of
Crater Flamsteed.

COMMAND AND SERVICE MODULE STRUCTURE, SYSTEMS

The Apollo spacecraft for the Apollo 11 mission is comprised of Command Module 107, Service Module 107, Lunar Module 5, a spacecraft-lunar module adapter (SLA) and a launch escape system. The SLA serves as a mating structure between the instrument unit atop the S-IVB stage of the Saturn V launch vehicle and as a housing for the lunar module.

Launch Escape System (LES) -- Propels command module to safety in an aborted launch. It is made up of an open-frame tower structure, mounted to the command module by four frangible bolts, and three solid-propellant rocket motors: a 147,000 pound-thrust launch excape system motor, a 2,400-pound-thrust pitch control motor, and a 31,500-pound-thrust tower jettison motor. Two canard vanes near the top deploy to turn the command module aerodynamically to an attitude with the heat-fhield forward. Attached to the base of the launch escape tower is a boost protective cover composed of resin impregnated fiberglass covered with cork, that protects the command module from aerodynamic heating during boost and rocket exhaust gases from the main and the jettison motors. The system is 33 feet tall, four feet in diameter at the base, and weighs 8,910 pounds.

Command Module (CM) Structure -- The basic structure of the command module is a pressure vessel encased in heat shields, cone-shaped 11 feet 5 inches high, base diameter of 12 feet 10 inches, and launch weight 12,250 pounds.

The command module consists of the forward compartment which contains two reaction control engines and components of the Earth landing system; the crew compartment or inner pressure vessel containing crew accomodations, controls and displays, and many of the spacecraft systems; and the aft compartment housing ten reaction control engines, propellant tankage, helium tanks, water tanks, and the CSM umbilical cable. The crew compartment contains 210 cubic feet of habitable volume.

Heat-shields around the three compartments are made of brazed stainless steel honeycomb with an outer layer of phenolic epoxy resin as an ablative material. Shield thickness, varying according to heat loads, ranges from 0.7 inch at the apex to 2.7 inches at the aft end.

The spacecraft inner structure is of sheet-aluminum honey-comb bonded sandwhich ranging in thickness from 0.25 inch thick at forward access tunnel to 1.5 inches thick at base.

APOLLO SPACECRAFT

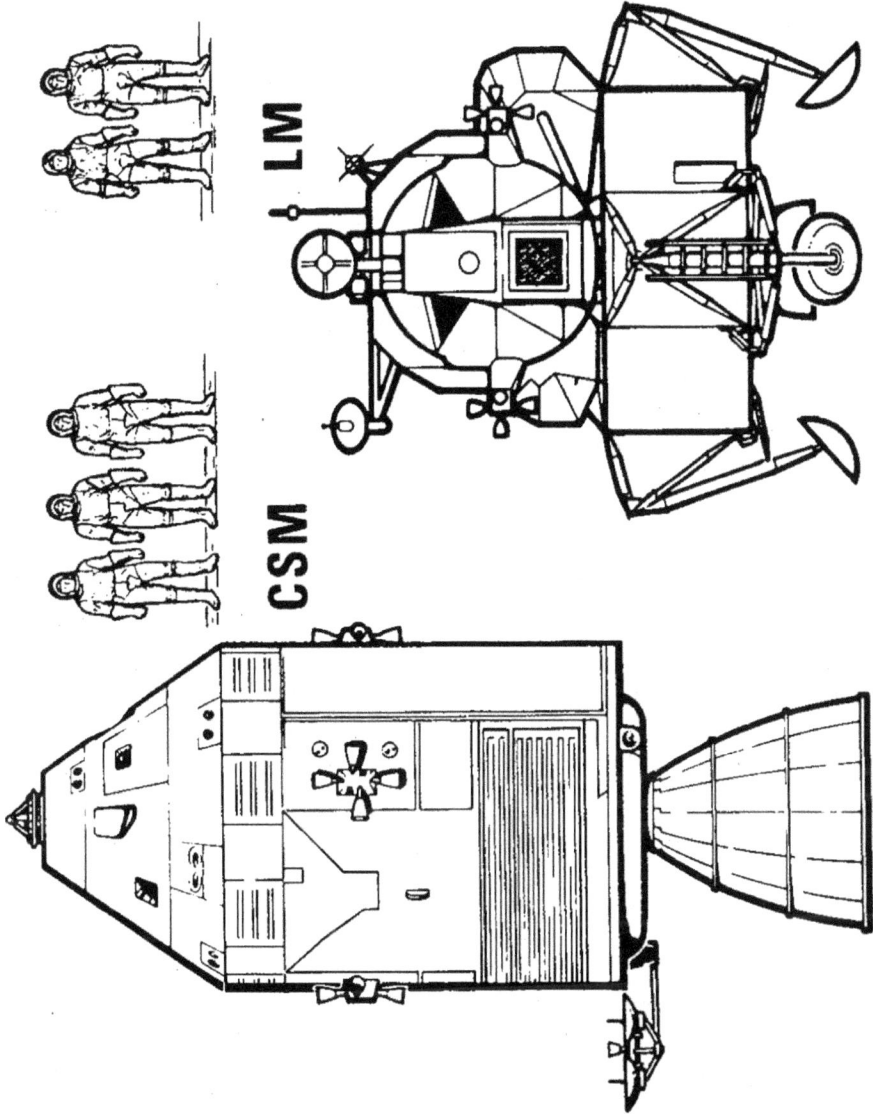

LM

CSM

CSM 107 and LM-5 are equipped with the probe-and-drogue docking hardware. The probe assembly is a powered folding coupling and impact attentuating device mounted on the CM tunnel that mates with a conical drogue mounted in the LM docking tunnel. After the 12 automatic docking latches are checked following a docking maneuver, both the probe and drogue assemblies are removed from the vehicle tunnels and stowed to allow free crew transfer between the CSM and LM.

Service Module (SM) Strucutre -- The service module is a cylinder 12 feet 10 inches in diameter by 24 feet 7 inches high. For the Apollo 11 mission, it will weigh, 51,243 pounds at launch. Aluminum honeycomb panels one inch thick form the outer skin, and milled aluminum radial beams separate the interior into six sections around a central cylinder containing two helium spheres, four sections containing service propulsion system fuel-oxidizer tankage, another containing fuel cells, cryogenic oxygen and hydrogen, and one sector essentially empty.

Spacecraft-LM Adapter (SLA) Structure -- The spacecraft LM adapter is a truncated cone 28 feet long tapering from 260 inches diameter at the base to 154 inches at the forward end at the service module mating line. Aluminum honeycomb 1.75 inches thick is the stressed-skin structure for the spacecraft adapter. The SLA weighs 4,000 pounds.

CSM Systems

Guidance, Navigation and Control System (GNCS) -- Measures and controls spacecraft position, attitude, and velocity, calculates trajectory, controls spacecraft propulsion system thrust vector, and displays abort data. The guidance system consists of three subsystems: inertial, made up of an inertial measurement unit and associated power and data components; computer which processes information to or from other components; and optics, including scanning telecope and sextant for celestial and/or landmark spacecraft navigation. CSM 107 and subsequent modules are equipped with a VHF ranging device as a backup to the LM rendezvous radar.

Stabilization and Control Systems (SCS) -- Controls spacecraft rotation, translation, and thrust vector and provides displays for crew-initiated maneuvers; backs up the guidance system. It has three subsystems; attitude reference, attitude control, and thrust vector control.

Service Propulsion System (SPS) -- Provides thrust for large spacecraft velocity changes through a gimbal-mounted 20,500-pound-thrust hypergolic engine using a nitrogen tetroxide oxidizer and a 50-50 mixture of unsymmetrical dimethyl hydrazine and hydrazine fuel. This system is in the service module. The system responds to automatic firing commands from the guidance and navigation system or to manual commands from the crew. The engine provides a constant thrust level. The stabilization and control system gimbals the engine to direct the thrust vector through the spacecraft center of gravity.

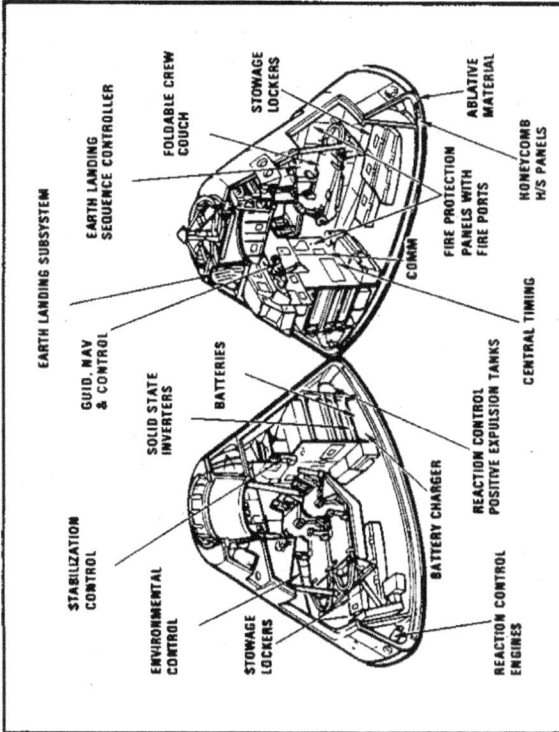

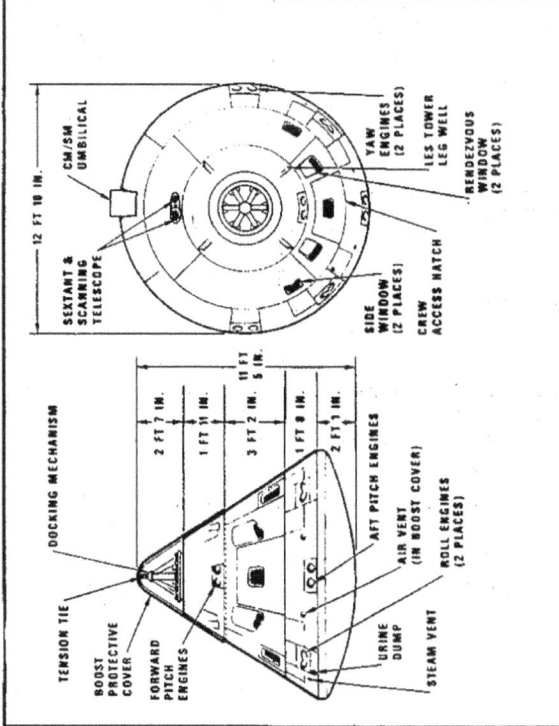

COMMAND MODULE

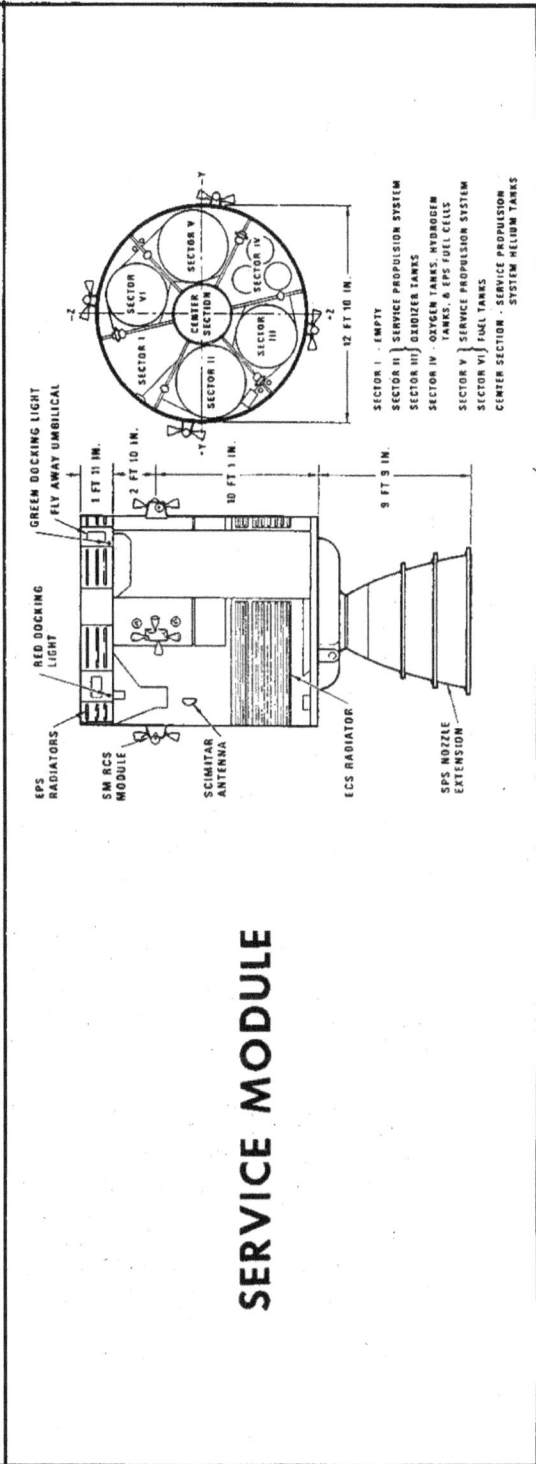

SERVICE MODULE

Telecommunications System -- Provides voice, television, telemetry, and command data and tracking and ranging between the spacecraft and Earth, between the command module and the lunar module
and between the spacecraft and the extravehicular astronaut. It
also provides intercommunications between astronauts. The telecommunications system consists of pulse code modulated telemetry
for relaying to Manned Space Flight Network stations data on spacecraft systems and crew condition, VHF/AM voice, and unified S-Band
tracking transponder, air-to-ground voice communications, onboard
television, and a VHF recovery beacon. Network stations can transmit
to the spacecraft such items as updates to the Apollo guidance
computer and central timing equipment, and real-time commands for
certain onboard functions.

The high-gain steerable S-Band antenna consists of four,
31-inch-diameter parabolic dishes mounted on a folding boom at
the aft end of the service module. Nested alongside the service
propulsion system engine nozzle until deployment, the antenna
swings out at right angles to the spacecraft longitudinal axis,
with the boom pointing 52 degrees below the heads-up horizontal.
Signals from the ground stations can be tracked either automatically
or manually with the antenna's gimballing system. Normal S-Band
voice and uplink/downlink communications will be handled by the
omni and high-gain antennas.

Sequential System -- Interfaces with other spacecraft systems
and subsystems to initiate time critical functions during launch,
docking maneuvers, sub-orbital aborts, and entry portions of a
mission. The system also controls routine spacecraft sequencing
such as service module separation and deployment of the Earth landing system.

Emergency Detection System (EDS) -- Detects and displays to
the crew launch vehicle emergency conditions, such as excessive
pitch or roll rates or two engines out, and automatically or
manually shuts down the booster and activates the launch escape
system; functions until the spacecraft is in orbit.

Earth Landing System (ELS) -- Includes the drogue and main
parachute system as well as post-landing recovery aids. In a
normal entry descent, the command module forward heat shield
is jettisoned at 24,000 feet, permitting mortar deployment of
two reefed 16.5-foot diameter drogue parachutes for orienting
and decelerating the spacecraft. After disreef and drogue release,
three mortar deployed pilot chutes pull out the three main 83.3-
foot diameter parachutes with two-stage reefing to provide gradual
inflation in three steps. Two main parachutes out of three can
provide a safe landing.

SPACECRAFT AXIS AND ANTENNA LOCATIONS

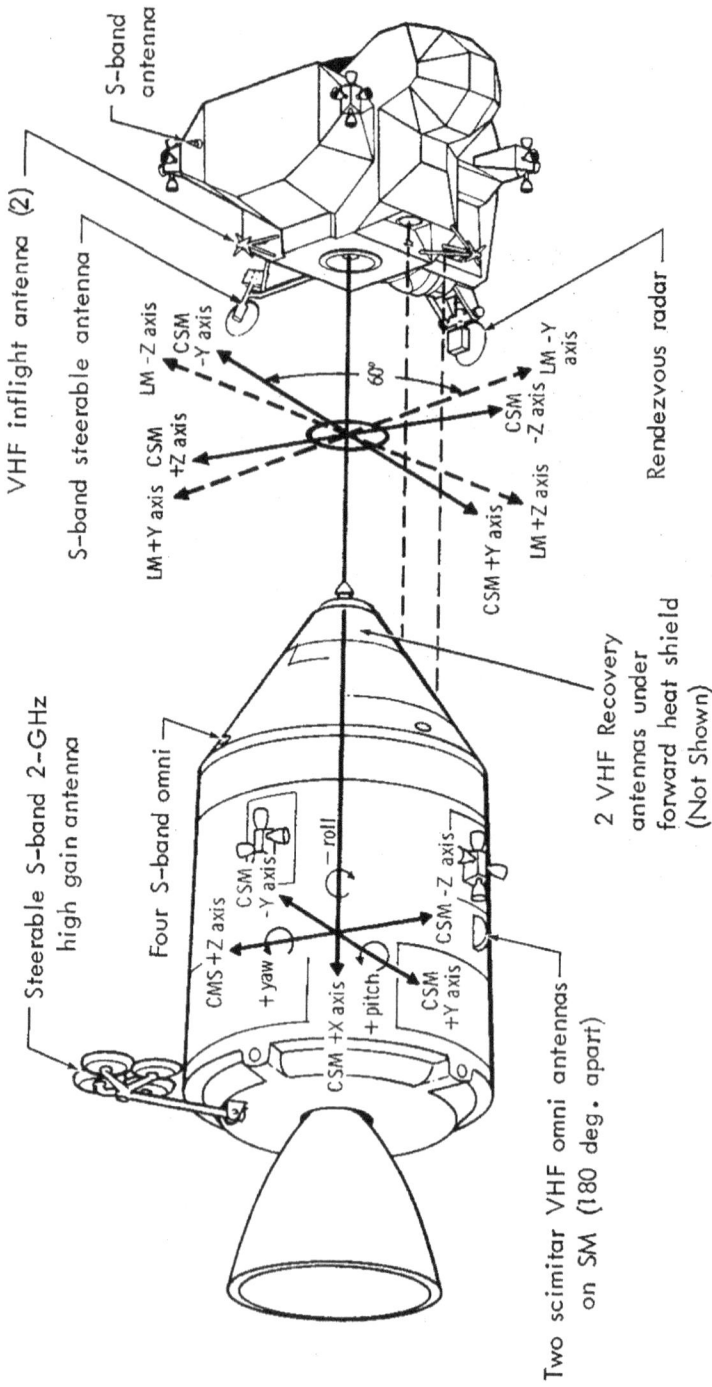

SPACECRAFT AXIS AND ANTENNA LOCATIONS

Reaction Control System (RCS) -- The command module and the service module each has its own independent system. The SM RCS has four identical RCS "quads" mounted around the SM 90 degrees apart. Each quad has four 100 pound-thrust engines, two fuel and two oxidizer tanks and a helium pressurization sphere. The SM RCS provides redundant spacecraft attitude control through cross-coupling logic inputs from the stabilization and guidance systems. Small velocity change maneuvers can also be made with the SM RCS.

The CM RCS consists of two independent six-engine subsystems of six 93 pound-thrust engines each. Both subsystems are activated just prior to CM separation from the SM: one is used for spacecraft attitude control during entry. The other serves in standby as a backup. Propellants for both CM and SM RCS are monomethyl hydrazine fuel and nitrogen tetroxide oxidizer with helium pressurization. These propellants are hypergolic, i.e., they burn spontaneously when combined without an igniter.

Electrical Power System (EPS) -- Provides electrical energy sources, power generation and control, power conversion and conditioning, and power distribution to the spacecraft throughout the mission. The EPS also furnishes drinking water to the astronauts as a by-product of the fuel cells. The primary source of electrical power is the fuel cells mounted in the SM. Each cell consists of a hydrogen compartment, an oxygen compartment, and two electrodes. The cryogenic gas storage system, also located in the SM, supplies the hydrogen and oxygen used in the fuel cell power plants, as well as the oxygen used in the ECS.

Three silver-zinc oxide storage batteries supply power to the CM during entry and after landing, provide power for sequence controllers, and supplement the fuel cells during periods of peak power demand. These batteries are located in the CM lower equipment bay. A battery charger is located in the same bay to assure a full charge prior to entry.

Two other silver-zinc oxide batteries, independent of and completely isolated from the rest of the dc power system, are used to supply power for explosive devices for CM/SM separation, parachute deployment and separation, third-stage separation, launch escape system tower separation, and other pyrotechnic uses.

Environmental Control System (ECS) -- Controls spacecraft atmosphere, pressure, and temperature and manages water. In addition to regulating cabin and suit gas pressure, temperature and humidity, the system removes carbon dioxide, odors and particles, and ventilates the cabin after landing. It collects and stores fuel cell potable water for crew use, supplies water to the glycol evaporators for cooling, and dumps surplus water overboard through the urine dump valve. Proper operating temperature of electronics and electrical equipment is maintained by this system through the use of the cabin heat exchangers, the space radiators, and the glycol evaporators.

Recovery aids include the uprighting system, swimmer inter-phone connections, sea dye marker, flashing beacon, VHF recovery beacon, and VHF transceiver. The uprighting system consists of three compressor-inflated bags to upright the spacecraft if it should land in the water apex down (stable II position).

Caution and Warning System -- Monitors spacecraft systems for out-of-tolerance conditions and alerts crew by visual and audible alarms so that crewmen may trouble-shoot the problem.

Controls and Displays -- Provide readouts and control functions of all other spacecraft systems in the command and service modules. All controls are designed to be operated by crewmen in pressurized suits. Displays are grouped by system and located according to the frequency the crew refers to them.

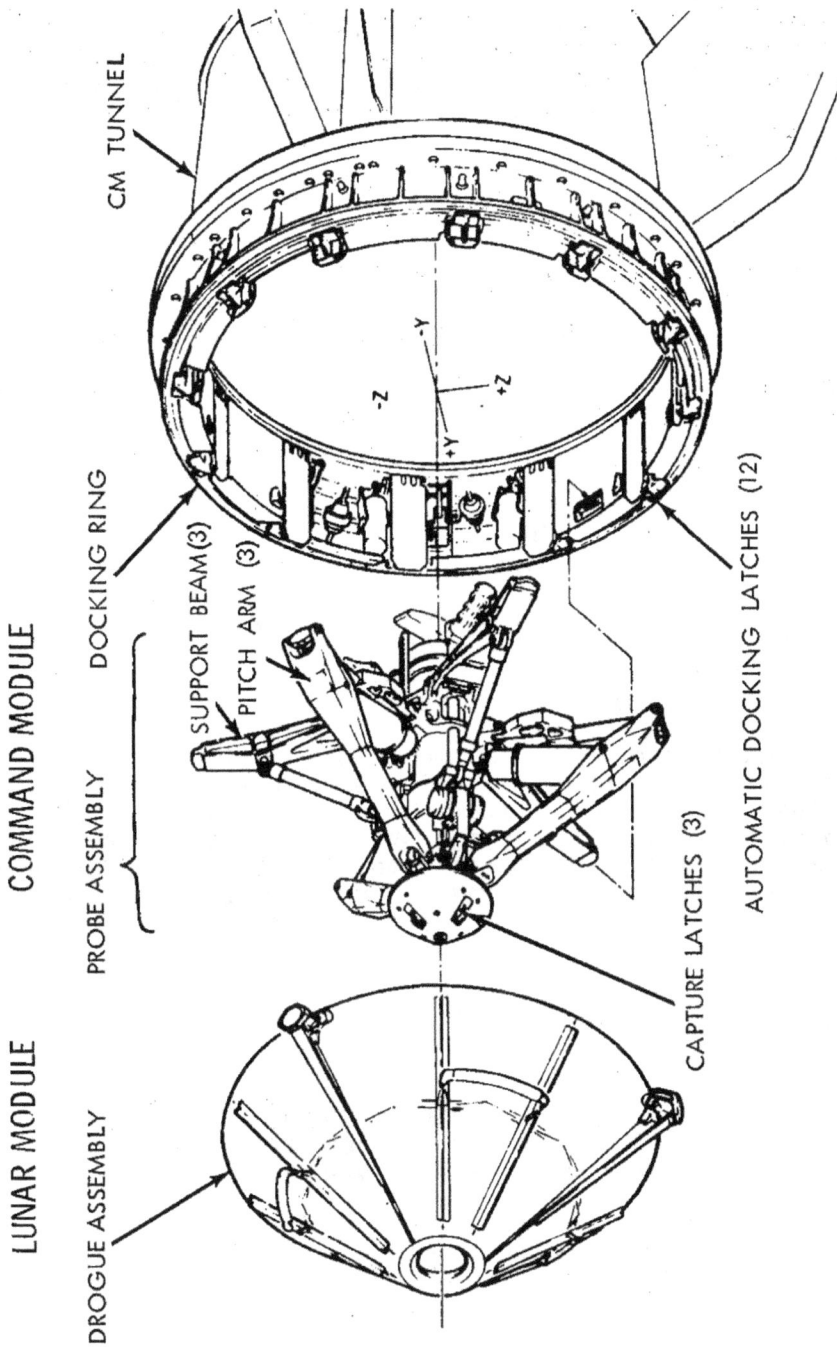

APOLLO DOCKING MECHANISMS

LUNAR MODULE

DROGUE ASSEMBLY

COMMAND MODULE

PROBE ASSEMBLY

SUPPORT BEAM (3)

PITCH ARM (3)

CAPTURE LATCHES (3)

AUTOMATIC DOCKING LATCHES (12)

DOCKING RING

CM TUNNEL

LUNAR MODULE STRUCTURES, WEIGHT

The lunar module is a two-stage vehicle designed for space operations near and on the Moon. The LM is incapable of reentering the atmosphere. The lunar module stands 22 feet 11 inches high and is 31 feet wide (diagonally across landing gear).

Joined by four explosive bolts and umbilicals, the ascent and descent stages of the LM operate as a unit until staging, when the ascent stage functions as a single spacecraft for rendezvous and docking with the CSM.

Ascent Stage

Three main sections make up the ascent stage: the crew compartment, midsection, and aft equipment bay. Only the crew compartment and midsection are pressurized (4.8 psig; 337.4 gm/sq cm) as part of the LM cabin; all other sections of the LM are unpressurized. The cabin volume is 235 cubic feet (6.7 cubic meters). The ascent stage measures 12 feet 4 inches high by 14 feet 1 inch in diameter.

Structurally, the ascent stage has six substructural areas: crew compartment, midsection, aft equipment bay, thrust chamber assembly cluster supports, antenna supports and thermal and micrometeoroid shield.

The cylindrical crew compartment is a semimonocoque structure of machined longerons and fusion-welded aluminum sheet and is 92 inches (2.35 m) in diameter and 42 inches (1.07 m) deep. Two flight stations are equipped with control and display panels, armrests, body restraints, landing aids, two front windows, an overhead docking window, and an alignment optical telescope in the center between the two flight stations. The habitable volume is 160 cubic feet.

Two triangular front windows and the 32-inch (0.81 m) square inward-opening forward hatch are in the crew compartment front face.

External structural beams support the crew compartment and serve to support the lower interstage mounts at their lower ends. Ring-stiffened semimonocoque construction is employed in the midsection, with chem-milled aluminum skin over fusion-welded longerons and stiffeners. Fore-and-aft beams across the top of the midsection join with those running across the top of the cabin to take all ascent stage stress loads and, in effect, isolate the cabin from stresses.

-more-

DOCKING WINDOW

DOCKING
DROGUE
ASSEMBLY

VHF ANTENNA

DOCKING
TARGET

S-BAND
STEERABLE
ANTENNA

EVA ANTENNA

RENDEZVOUS
RADAR ANTENNA

AFT
EQUIPMENT
BAY

S-BAND IN-FLIGHT
ANTENNA (2)

RCS THRUST
CHAMBER
ASSEMBLY
CLUSTER (4)

WINDOWS (2)

DOCKING
LIGHT (4)

LANDING
GEAR

TRACKING LIGHT

LANDING
PAD

FORWARD
HATCH

FORWARD

+Z

LADDER

EGRESS
PLATFORM

DESCENT
ENGINE
SKIRT

LANDING RADAR
ANTENNA

LUNAR SURFACE SENSING PROBE (3)

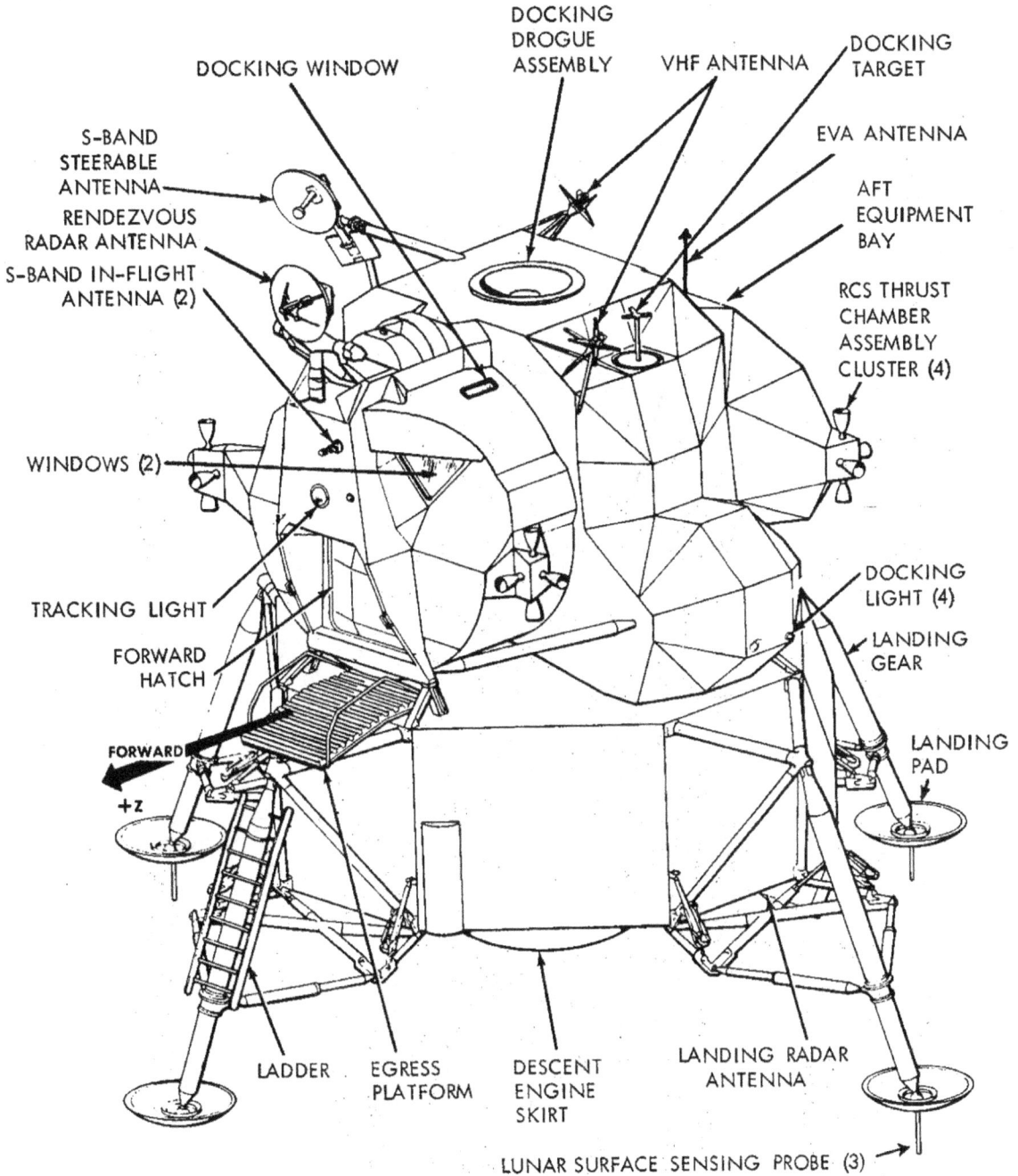

APOLLO LUNAR MODULE

-more-

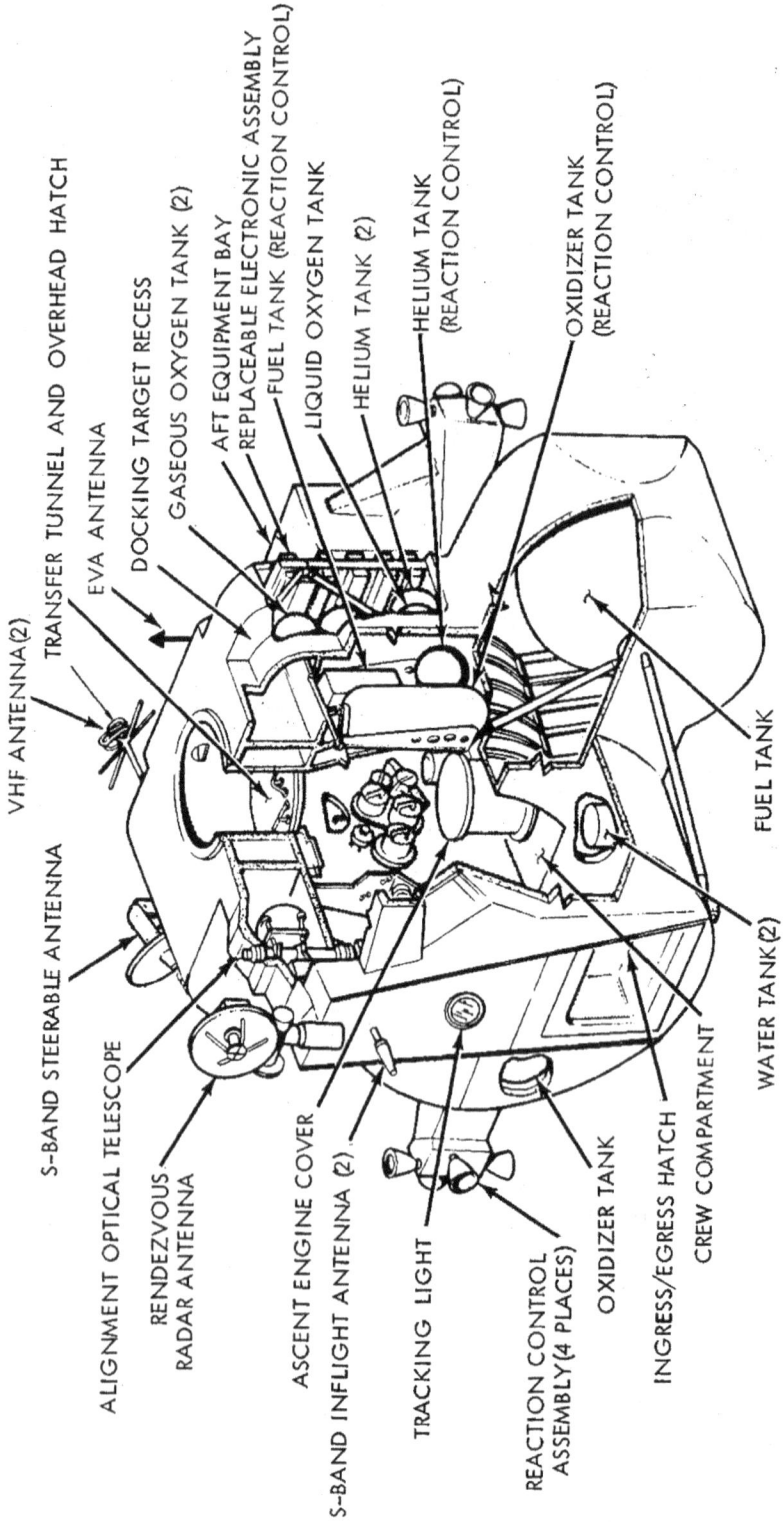

APOLLO LUNAR MODULE - ASCENT STAGE

ECS LIOH CARTRIDGE

ECS CREW UMBILICALS

ENVIRONMENTAL CONTROL SUBSYSTEM

ALIGNMENT
OPTICAL
TELESCOPE

PLSS
RECHARGE HOSE

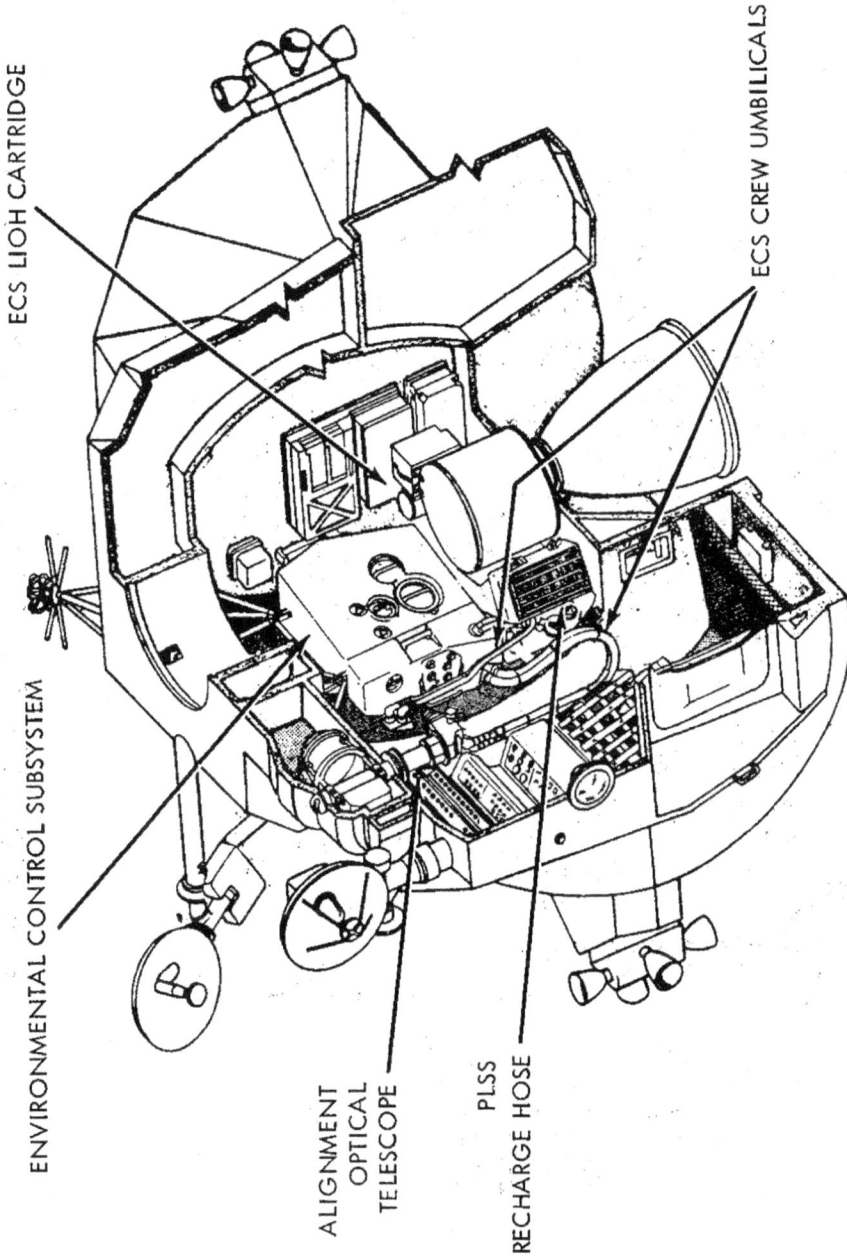

LM CABIN INTERIOR, LEFT HALF

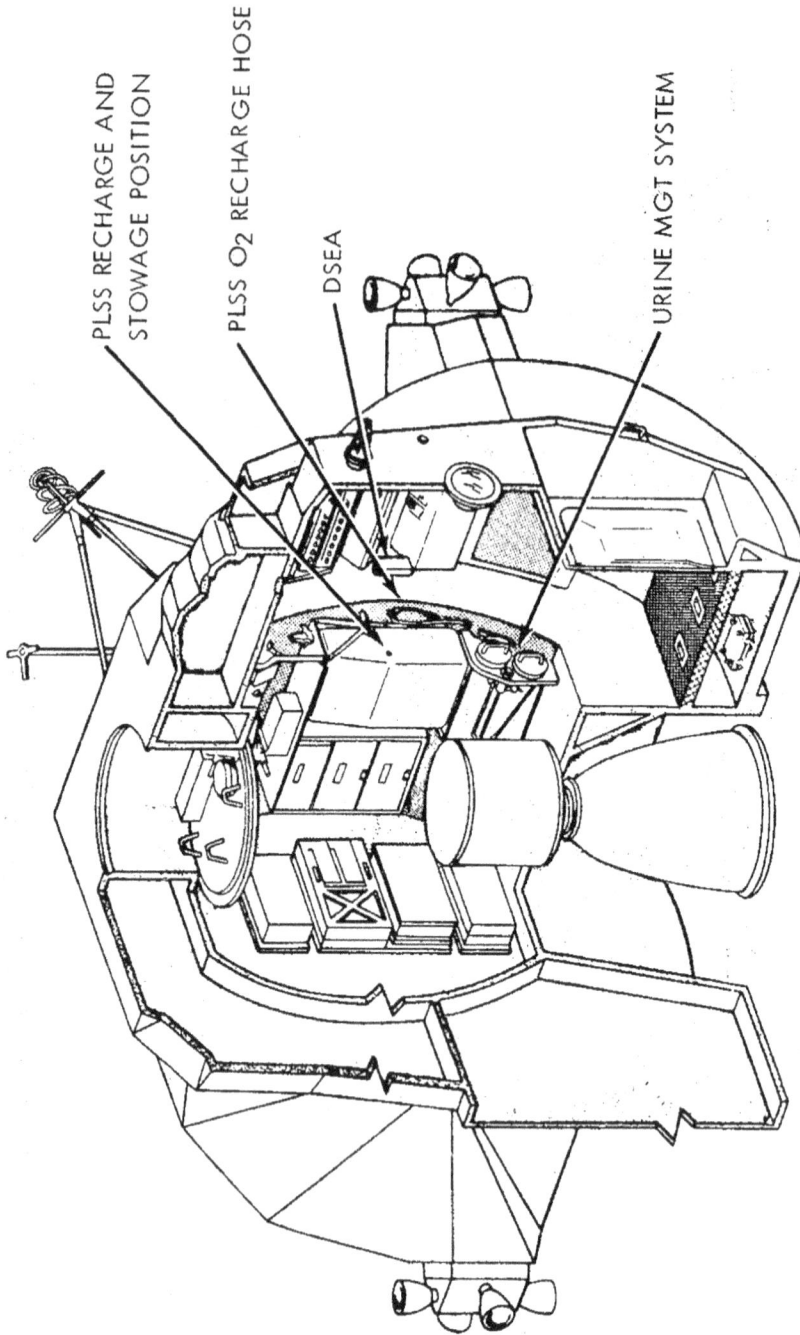

PLSS RECHARGE AND STOWAGE POSITION

PLSS O2 RECHARGE HOSE

DSEA

URINE MGT SYSTEM

LM CABIN INTERIOR, RIGHT HALF

The ascent stage engine compartment is formed by two beams running across the lower midsection deck and mated to the fore and aft bulkheads. Systems located in the midsection include the LM guidance computer, the power and servo assembly, ascent engine propellant tanks, RCS propellant tanks, the environmental control system, and the waste management section.

A tunnel ring atop the ascent stage meshes with the command module docking latch assemblies. During docking, the CM docking ring and latches are aligned by the LM drogue and the CSM probe.

The docking tunnel extends downward into the midsection 16 inches (40 cm). The tunnel is 32 inches (0.81 cm) in diameter and is used for crew transfer between the CSM and LM. The upper hatch on the inboard end of the docking tunnel hinges downward and cannot be opened with the LM pressurized and undocked.

A thermal and micrometeoroid shield of multiple layers of mylar and a single thickness of thin aluminum skin encases the entire ascent stage structure.

Descent Stage

The descent stage consists of a cruciform load-carrying structure of two pairs of parallel beams, upper and lower decks, and enclosure bulkheads -- all of conventional skin-and-stringer aluminum alloy construction. The center compartment houses the descent engine, and descent propellant tanks are housed in the four square bays around the engine. The descent stage measures 10 feet 7 inches high by 14 feet 1 inch in diameter.

Four-legged truss outriggers mounted on the ends of each pair of beams serve as SLA attach points and as "knees" for the landing gear main struts.

Triangular bays between the main beams are enclosed into quadrants housing such components as the ECS water tank, helium tanks, descent engine control assembly of the guidance, navigation and control subsystem, ECS gaseous oxygen tank, and batteries for the electrical power system. Like the ascent stage, the descent stage is encased in the mylar and aluminum alloy thermal and micrometeoroid shield.

The LM external platform, or "porch", is mounted on the forward outrigger just below the forward hatch. A ladder extends down the forward landing gear strut from the porch for crew lunar surface operations.

-more-

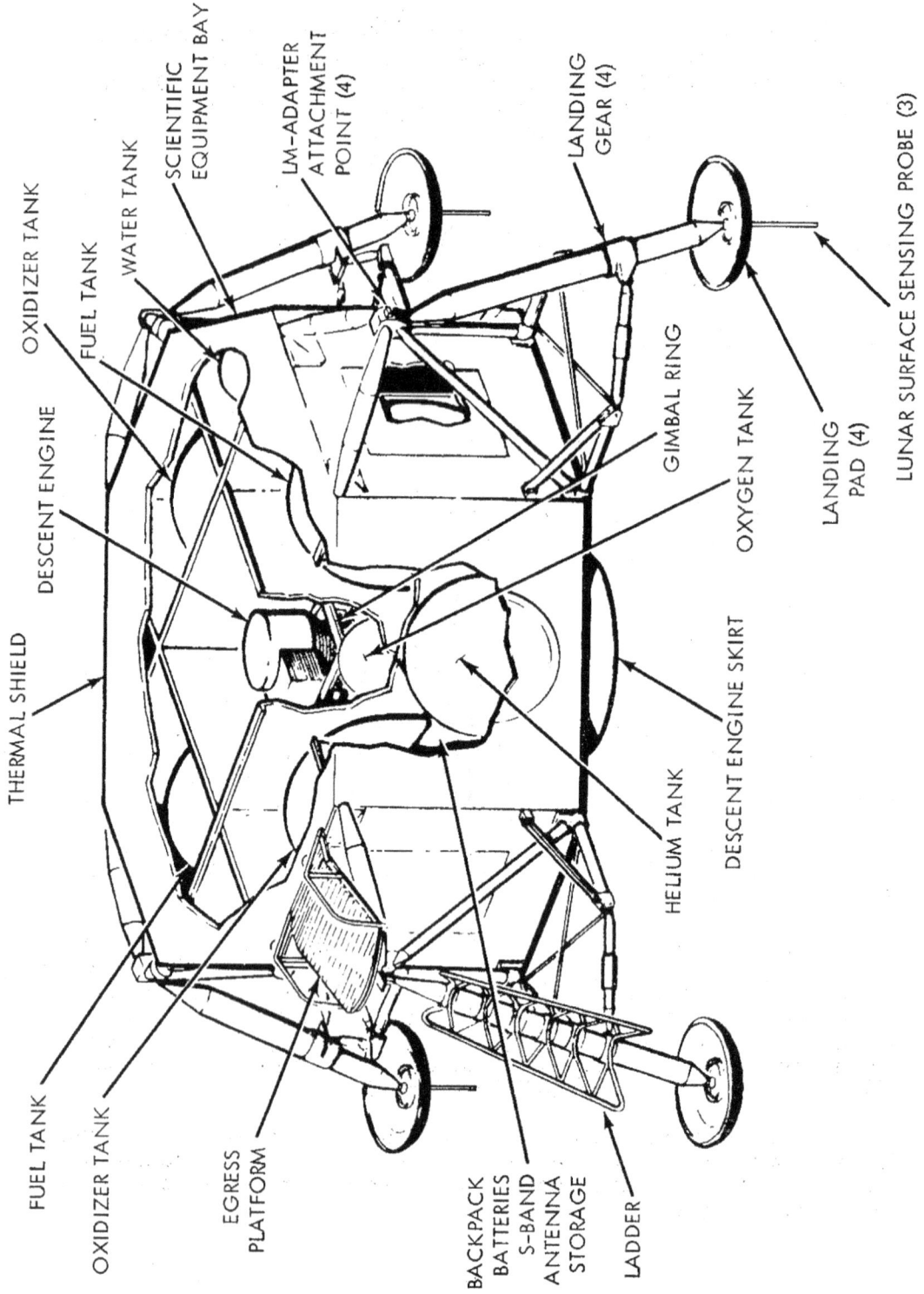

In a retracted position until after the crew mans the
LM, the landing gear struts are explosively extended and
provide lunar surface landing impact attenuation. The main
struts are filled with crushable aluminum honeycomb for
absorbing compression loads. Footpads 37 inches (0.95 m) in
diameter at the end of each landing gear provide vehicle
"floatation" on the lunar surface.

Each pad (except forward pad) is fitted with a lunar-
surface sensing probe which signals the crew to shut down
the descent engine upon contact with the lunar surface.

LM-5 flown on the Apollo 11 mission will have a launch
weight of 33,205 pounds. The weight breakdown is as follows:

Ascent stage, dry	4,804 lbs.	Includes water and oxygen; no crew
Descent stage, dry	4,483 lbs.	
RCS propellants (loaded)	604 lbs.	
DPS propellants (loaded)	18,100 lbs.	
APS propellants (loaded)	5,214 lbs.	
	33,205 lbs.	

Lunar Module Systems

Electrical Power System -- The LM DC electrical system
consists of six silver zinc primary batteries -- four in the
descent stage and two in the ascent stage, each with its own
electrical control assembly (ECA). Power feeders from all
primary batteries pass through circuit breakers to energize
the LM DC buses, from which 28-volt DC power id distributed
through circuit breakers to all LM systems. AC power
(117v 400Hz) is supplied by two inverters, either of which can
supply spacecraft AC load needs to the AC buses.

Environmental Control System -- Consists of the atmosphere
revitalization section, oxygen supply and cabin pressure control
section, water management, heat transport section, and outlets
for oxygen and water servicing of the Portable Life Support
System (PLSS).

Components of the atmosphere revitalization section are the suit circuit assembly which cools and ventilates the pressure garments, reduces carbon dioxide levels, removes odors, noxious gases and excessive moisture; the cabin recirculation assembly which ventilates and controls cabin atmosphere temperatures; and the steam flex duct which vents to space steam from the suit circuit water evaporator.

The oxygen supply and cabin pressure section supplies gaseous oxygen to the atmosphere revitalization section for maintaining suit and cabin pressure. The descent stage oxygen supply provides descent flight phase and lunar stay oxygen needs, and the ascent stage oxygen supply provides oxygen needs for the ascent and rendezvous flight phase.

Water for drinking, cooling, fire fighting, food preparation, and refilling the PLSS cooling water servicing tank is supplied by the water management section. The water is contained in three nitrogen-pressurized bladder-type tanks, one of 367-pound capacity in the descent stage and two of 47.5-pound capacity in the ascent stage.

The heat transport section has primary and secondary water-glycol solution coolant loops. The primary coolant loop circulates water-glycol for temperature control of cabin and suit circuit oxygen and for thermal control of batteries and electronic components mounted on cold plates and rails. If the primary loop becomes inoperative, the secondary loop circulates coolant through the rails and cold plates only. Suit circuit cooling during secondary coolant loop operation is provided by the suit loop water boiler. Waste heat from both loops is vented overboard by water evaporation or sublimators.

Communication System -- Two S-band transmitter-receivers, two VHF transmitter-receivers, a signal processing assembly, and associated spacecraft antenna make up the LM communications system. The system transmits and receives voice, tracking and ranging data, and transmits telemetry data on about 270 measurements and TV signals to the ground. Voice communications between the LM and ground stations is by S-band, and between the LM and CSM voice is on VHF.

-more-

Although no real-time commands can be sent to LM-5 and subsequent spacecraft, the digital uplink is retained to process guidance officer commands transmitted from Mission Control Center to the LM guidance computer, such as state vector updates.

The data storage electronics assembly (DSEA) is a four-channel voice recorder with timing signals with a 10-hour recording capacity which will be brough back into the CSM for return to Earth. DSEA recordings cannot be "dumped" to ground stations.

LM antennas are one 26-inch diameter parabolic S-band steerable antenna, two S-band inflight antennas, two VHF inflight antennas, and an erectable S-band antenna (optional) for lunar surface.

Guidance, Navigation and Control System -- Comprised of six sections: primary guidance and navigation section (PGNS), abort guidance section (AGS), radar section, control electronics section (CES), and orbital rate drive electronics for Apollo and LM (ORDEAL).

* The PGNS is an aided inertial guidance system updated by the alignment optical telescope, an inertial measurement unit, and the rendezvous and landing radars. The system provides inertial reference data for computations, produces inertial alignment reference by feeding optical sighting data into the LM guidance computer, displays position and velocity data, computes LM-CSM rendezvous data from radar inputs, controls attitude and thrust to maintain desired LM trajectory, and controls descent engine throttling and gimbaling.

The LM-5 guidance computer has the Luminary IA software program for processing landing radar altitude and velocity information for lunar landing. LM-4, flown on Apollo 10, did not have the landing phase in its guidance computer Luminary I program.

* The AGS is an independent backup system for the PGNS, having its own inertial sensors and computer.

* The radar section is made up of the rendezvous radar which provides CSM range and range rate, and line-of-sight angles for maneuver computation to the LM guidance computer; the landing radar which provide altitude and velocity data to the LM guidance computer during lunar landing. The rendezvous radar has an operating range from 80 feet to 400 nautical miles. The range transfer tone assembly, utilizing VHF electronics, is a passive responder to the CSM VHF ranging device and is a backup to the rendezvous radar.

• The CES controls LM attitude and translation about all axes. It also controls by PGNS command the automatic operation of the ascent and descent engines, and the reaction control thrusters. Manual attitude controller and thrust-translation controller commands are also handled by the CES.

* ORDEAL, displays on the flight director attitude indicator, is the computed local vertical in the pitch axis during circular Earth or lunar orbits.

Reaction Control System -- The LM has four RCS engine clusters of four 100-pound (45.4 kg) thrust engines each which use helium-pressurized hypergolic propellants. The oxidizer is nitrogen tetroxide, fuel is Aerozine 50 (50/50 blend of hydrazine and unsymmetrical dimethyl hydrazine). Propellant plumbing, valves and pressurizing components are in two parallel, independent systems, each feeding half the engines in each cluster. Either system is capable of maintaining attitude alone, but if one supply system fails, a propellant crossfeed allows one system to supply all 16 engines. Additionally, interconnect valves permit the RCS system to draw from ascent engine propellant tanks.

The engine clusters are mounted on outriggers 90 degrees apart on the ascent stage.

The RCS provides small stabilizing impulses during ascent and descent burns, controls LM attitude during maneuvers, and produces thrust for separation, and ascent/descent engine tank ullage. The system may be operated in either the pulse or steady-state modes.

Descent Propulsion System -- Maximum rated thrust of the descent engine is 9,870 pounds (4,380.9 kg) and is throttleable between 1,050 pounds (476.7 kg) and 6,300 pounds (2,860.2 kg). The engine can be gimbaled six degrees in any direction in response to attitude commands and for offset center of gravity trimming. Propellants are helium-pressurized Aerozine 50 and nitrogen tetroxide.

Ascent Propulsion System -- The 3,500-pound (1,589 kg) thrust ascent engine is not gimbaled and performs at full thrust. The engine remains dormant until after the ascent stage separates from the descent stage. Propellants are the same as are burned by the RCS engines and the descent engine.

Caution and Warning, Controls and Displays -- These two systems have the same function aboard the lunar module as they do aboard the command module. (See CSM systems section.)

<u>Tracking and Docking Lights</u> -- A flashing tracking light (once per second, 20 milliseconds duration) on the front face of the lunar module is an aid for contingency CSM-active rendezvous LM rescue. Visibility ranges from 400 nautical miles through the CSM sextant to 130 miles with the naked eye. Five docking lights analagous to aircraft running lights are mounted on the LM for CSM-active rendezvous: two forward yellow lights, aft white light, port red light and starboard green light. All docking lights have about a 1,000-foot visibility.

SATURN V LAUNCH VEHICLE DESCRIPTION AND OPERATION

The Apollo 11 spacecraft will be boosted into Earth orbit and then onto a lunar trajectory by the sixth Saturn V launch vehicle. The 281-foot high Saturn V generates enough thrust to place a 125-ton payload into a 105 nm Earth orbit or boost about 50 tons to lunar orbit.

The Saturn V, developed by the NASA-Marshall Space Flight Center, underwent research and development testing in the "all-up" mode. From the first launch all stages have been live. This has resulted in "man rating" of the Saturn V in two launches. The third Saturn V (AS-503) carried Apollo 8 and its crew on a lunar orbit mission.

Saturn V rockets were launched November 9, 1967, April 4, 1968, December 21, 1968, March 3, 1969, and May 18, 1969. The first two space vehicle were unmanned; the last three carried the Apollo 8, 9 and 10 crews, respectively.

Launch Vehicle Range Safety Provisions

In the event of an imminent emergency during the launch vehicle powered flight phase it could become necessary to abort the mission and remove the command module and crew from immediate danger. After providing for crew safety, the Range Safety Officer may take further action if the remaining intact vehicle constitutes a hazard to overflown geographic areas. Each launch vehicle propulsive stage is equipped with a propellant dispersion system to terminate the vehicle flight in a safe location and disperse propellants with a minimized ignition probability. A transmitted ground command shuts down all engines and a second command detonates explosives which open the fuel and oxidizer tanks enabling the propellants to disperse. On each stage the tank cuts are made in non-adjacent areas to minimize propellant mixing. The stage propellant dispersion systems are safed by ground command.

SATURN V LAUNCH VEHICLE

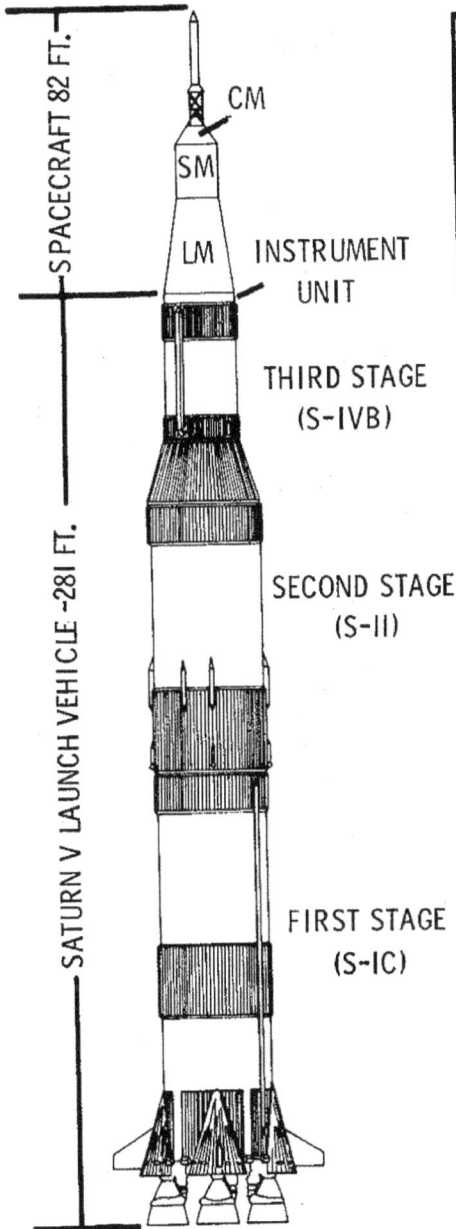

SPACECRAFT 82 FT.

CM

SM

LM

INSTRUMENT
UNIT

THIRD STAGE
(S-IVB)

SECOND STAGE
(S-II)

SATURN V LAUNCH VEHICLE -281 FT.

FIRST STAGE
(S-IC)

FIRST STAGE (S-IC)	
DIAMETER	33 FEET
HEIGHT	138 FEET
WEIGHT	5,022,674 LBS. FUELED
	288,750 LBS. DRY
ENGINES	FIVE F-I
PROPELLANTS	LIQUID OXYGEN (3,307,855 LBS.,
	346,372 GALS.) RP-I (KEROSENE)
	- (1,426,069 LBS., 212,846 GALS.)
THRUST	7,653,854 LBS. AT LIFTOFF

SECOND STAGE (S-II)	
DIAMETER	33 FEET
HEIGHT	81.5 FEET
WEIGHT	1,059,171 LBS. FUELED
	79,918 LBS. DRY
ENGINES	FIVE J-2
PROPELLANTS	LIQUID OXYGEN (821,022 LBS.,
	85,973 GALS.) LIQUID HYDROGEN
	(158,221 LBS., 282,555 GALS.)
THRUST	1,120,216 TO 1,157,707 LBS.
INTERSTAGE	1,353 (SMALL)
	8,750 (LARGE)

THIRD STAGE (S-IVB)	
DIAMETER	21.7 FEET
HEIGHT	58.3 FEET
WEIGHT	260,523 LBS. FUELED
	25,000 LBS. DRY
ENGINES	ONE J-2
PROPELLANTS	LIQUID OXYGEN (192,023 LBS.,
	20,107 GALS.) LIQUID HYDROGEN
	(43,500 LBS., 77,680 GALS.)
THRUST	178,161 TO 203,779 LBS.
INTERSTAGE	8,081 LBS.

INSTRUMENT UNIT	
DIAMETER	21.7 FEET
HEIGHT	3 FEET
WEIGHT	4,306 LBS.

NOTE: WEIGHTS AND MEASURES GIVEN ABOVE ARE FOR THE NOMINAL VEHICLE CONFIGURATION FOR APOLLO 11. THE FIGURES MAY VARY SLIGHTLY DUE TO CHANGES BEFORE LAUNCH TO MEET CHANGING CONDITIONS. WEIGHTS NOT INCLUDED IN ABOVE ARE FROST AND MISCELLANEOUS SMALLER ITEMS.

SPACE VEHICLE WEIGHT SUMMARY (pounds)

Event	Wt. Chg.	Veh. Wt.
At ignition		6,484,280
Thrust buildup propellant used	85,745	
At first motion		6,398,535
S-IC frost	650	
S-IC nitrogen purge	37	
S-II frost	450	
S-II insulation purge gas	120	
S-IVB frost	200	
Center engine decay propellant used	2,029	
Center engine expended propellant	406	
S-IC mainstage propellant used	4,567,690	
Outboard engine decay propellant used	8,084	
S-IC stage drop weight	363,425	
S-IC/S-II small interstage	1,353	
S-II ullage propellant used	73	
At S-IC separation		1,454,014
S-II thrust buildup propellant used	1,303	
S-II start tank	25	
S-II ullage propellant used	1,288	
S-II mainstage propellant and venting	963,913	
Launch escape tower	8,930	
S-II aft interstage	8,750	
S-II thrust decay propellant used	480	
S-II stage drop weight	94,140	
S-II/S-IVB interstage	8,081	
S-IVB aft frame dropped	48	
S-IVB detonator package	3	
At S-II/S-IVB separation		367,053
S-IVB ullage rocket propellant	96	
At S-IVB ignition		366,957
S-IVB ullage propellant	22	
S-IVB hydrogen in start tank	4	
Thrust buildup propellant	436	
S-IVB mainstage propellant used	66,796	
S-IVB ullage rocket cases	135	
S-IVB APS propellant	2	
At first S-IVB cutoff signal		299,586
Thrust decay propellant used	89	
APS propellant (ullage)	5	
Engine propellant lost	30	
At parking orbit insertion		299,562
Fuel tank vent	2,879	
APS propellant	235	
Hydrogen in start tank	2	
O2/H2 burner	16	
LOX tank vent	46	
S-IVB fuel lead loss	5	

Event	Wt. Chg.	Veh. Wt.
At second S-IVB ignition		296,379
S-IVB hydrogen in start tank	4	
Thrust buildup propellant	569	
S-IVB mainstage propellant used	164,431	
APS propellant used	8	
At second S-IVB cutoff signal		139,533
Thrust decay propellant used	124	
Engine propellant lost	40	
At translunar injection		139,369

. -more-

First Stage

The 7.6 million pound thrust first stage (S-IC) was developed jointly by the National Aeronautics and Space Administration's Marshall Space Flight Center and the Boeing Co.

The Marshall Center assembled four S-IC stages: a structural test model, a static test version, and the first two flight stages. Subsequent flight stages are assembled by Boeing at the Michoud Assembly Facility, New Orleans.

The S-IC for the Apollo 11 mission was the third flight booster tested at the NASA-Mississippi Test Facility. The first S-IC test at MTF was on May 11, 1967, the second on August 9, 1967, and the third--the booster for Apollo 11--was on August 6, 1968. Earlier flight stages were static fired at the Marshall Center.

The booster stage stands 138 feet high and is 33 feet in diameter. Major structural components include thrust structure, fuel tank, intertank structure, oxidizer tank, and forward skirt. Its five engines burn kerosene (RP-1) fuel and liquid oxygen. The stage weighs 288,750 empty and 5,022,674 pounds fueled.

Normal propellant flow rate to the five F-1 engines is 29,364.5 pounds (2,230 gallons) per second. Four of the engines are mounted on a ring, at 90 degree intervals. These four are gimballed to control the rocket's direction of flight. The fifth engine is mounted rigidly in the center.

Second Stage

The Space Division of North American Rockwell Corp. builds the 1 million pound thrust S-II stage at Seal Beach, California. The 81 foot 7 inch long, 33 foot diameter stage is made up of the forward skirt to which the third stage attaches, the liquid hydrogen tank, liquid oxygen tank (separated from the hydrogen tank by an insulted common bulkhead), the thrust structure on which the engines are mounted, and an interstage section to which the first stage attaches.

Five J-2 engines power the S-II. The outer four engines are equally spaced on a 17.5 foot diameter circle. These four engines may be gimballed through a plus or minus seven-degree square pattern for thrust vector control. As on the first stage, the center engine (number 5) is mounted on the stage centerline and is fixed in position.

The second stage (S-II), like the third stage, uses high performance J-2 engines that burn liquid oxygen and liquid hydrogen. The stage's purpose is to provide stage boost almost to Earth orbit.

The S-II for Apollo 11 was static tested by North American Rockwell at the NASA-Mississippi Test Facility on September 3, 1968. This stage was shipped to test site via the Panama Canal for the test firing.

Third Stage

The third stage (S-IVB) was developed by the McDonnell Douglas Astronautics Co. at Huntington Beach, Calif. At Sacramento, Calif., the stage passed a static firing test on July 17, 1968, as part of Apollo 11 mission preparation. The stage was flown directly to the NASA-Kennedy Space Center by the special aircraft, Super Guppy.

Measuring 58 feet 4 inches long and 21 feet 8 inches in diameter, the S-IVB weighs 25,000 pounds dry. At first ignition, it weighs 262,000 pounds. The interstage section weighs an additional 8,081 pounds.

The fuel tanks contain 43,500 pounds of liquid hydrogen and 192,023 pounds of liquid oxygen at first ignition, totalling 235,523 pounds of propellants. Insulation between the two tanks is necessary because the liquid oxygen, at about 293 degrees below zero Fahrenheit, is warm enough, relatively, to rapidly heat the liquid hydrogen, at 423 degrees below zero, and cause it to turn to gas. The single J-2 engine produces a maximum 230,000 pounds of thrust. The stage provides propulsion twice during the Apollo 11 mission.

Instrument Unit

The instrument unit (IU) is a cylinder three feet high and 21 feet 8 inches in diameter. It weighs 4,306 pounds and contains the guidance, navigation and control equipment to steer the vehicle through its Earth orbits and into the final translunar injection maneuver.

The IU also contains telemetry, communications, tracking, and crew safety systems, along with its own supporting electrical power and environmental control systems.

Components making up the "brain" of the Saturn V are mounted on cooling panels fastened to the inside surface of the instrument unit skin. The "cold plates" are part of a system that removes heat by circulating cooled fluid through a heat exchanger that evaporates water from a separate supply into the vacuum of space.

The six major systems of the instrument unit are structural, thermal control, guidance and control, measuring and telemetry, radio frequency, and electrical.

The instrument unit provides navigation, guidance, and control of the vehicle· measurement of the vehicle performance and environment; data transmission with ground stations; radio tracking of the vehicle; checkout and monitoring of vehicle functions; initiation of stage functional sequencing; detection of emergency situations; generation and network distribution of electric power system operation; and preflight checkout and launch and flight operations.

A path-adaptive guidance scheme is used in the Saturn V instrument unit. A programmed trajectory is used during first stage boost with guidance beginning only after the vehicle has left the atmosphere. This is to prevent movements that might cause the vehicle to break apart while attempting to compensate for winds, jet streams, and gusts encountered in the atmosphere.

If after second stage ignition the vehicle deviates from the optimum trajectory in climb, the vehicle derives and corrects to a new trajectory. Calculations are made about once each second throughout the flight. The launch vehicle digital computer and data adapter perform the navigation and guidance computations and the flight control computer converts generated attitude errors into control commands.

The ST-124M inertial platform--the heart of the navigation, guidance and control system--provides space-fixed reference coordinates and measures acceleration along the three mutually perpendicular axes of the coordinate system. If the inertial platform fails during boost, spacecraft systems continue guidance and control functions for the rocket. After second stage ignition the crew can manually steer the space vehicle.

International Business Machines Corp., is prime contractor for the instrument unit and is the supplier of the guidance signal processor and guidance computer. Major suppliers of instrument unit components are: Electronic Communications, Inc., control computer; Bendix Corp., ST-124M inertial platform; and IBM Federal Systems Division, launch vehicle digital computer and launch vehicle data adapter.

Propulsion

The 41 rocket engines of the Saturn V have thrust ratings ranging from 72 pounds to more than 1.5 million pounds. Some engines burn liquid propellants, others use solids.

The five F-1 engines in the first stage burn RP-1 (kerosene) and liquid oxygen. Engines in the first stage develop approximately 1,530,771 pounds of thrust each at liftoff, building up to about 1,817,684 pounds before cutoff. The cluster of five engines gives the first stage a thrust range of from 7,653,854 pounds at liftoff to 9,088,419 pounds just before center engine cutoff.

The F-1 engine weighs almost 10 tons, is more than 18 feet high and has a nozzle-exit diameter of nearly 14 feet. The F-1 undergoes static testing for an average 650 seconds in qualifying for the 160-second run during the Saturn V first stage booster phase. The engine consumes almost three tons of propellants per second.

The first stage of the Saturn V for this mission has eight other rocket motors. These are the solid-fuel retrorockets which will slow and separate the stage from the second stage. Each rocket produces a thrust of 87,900 pounds for 0.6 second.

The main propulsion for the second stage is a cluster of five J-2 engines burning liquid hydrogen and liquid oxygen. Each engine develops a mean thrust of more than 227,000 pounds at 5:1 mixture ratio (variable from 224,000 to 231,000 in phases of this flight), giving the stage a total mean thrust of more than 1.135 million pounds.

Designed to operate in the hard vacuum of space, the 3,500-pound J-2 is more efficient than the F-1 because it burns the high-energy fuel hydrogen. F-1 and J-2 engines are produced by the Rocketdyne Division of North American Rockwell Corp.

The second stage has four 21,000-pound-thrust solid fuel rocket engines. These are the ullage rockets mounted on the S-IC/S-II interstage section. These rockets fire to settle liquid propellant in the bottom of the main tanks and help attain a "clean" separation from the first stage; they remain with the interstage when it drops away at second plane separation. Four retrorockets are located in the S-IVB aft interstage (which never separates from the S-II) to separate the S-II from the S-IVB prior to S-IVB ignition.

Eleven rocket engines perform various functions on the third stage. A single J-2 provides the main propulsive force; there are two jettisonable main ullage rockets and eight smaller engines in the two auxiliary propulsion system modules.

Launch Vehicle Instrumentation and Communication

A total of 1,348 measurements will be taken in flight on the Saturn V launch vehicle: 330 on the first stage, 514 on the second stage, 283 on the third stage, and 221 on the instrument unit.

Telemetry on the Saturn V includes FM and PCM systems on the S-IC, two FM and a PCM on the S-II, a PCM on the S-IVB, and an FM, a PCM and a CCS on the IU. Each propulsive stage has a range safety system, and the IU has C-Band and command systems.

Note:
FM (Frequency Modulated) PCM (Pulse Code Modulated) CCS (Command Communications System)

S-IVB Restart

The third stage of the Saturn V rocket for the Apollo mission will burn twice in space. The second burn places the spacecraft on the translunar trajectory. The first opportunity for this burn is at 2 hours 44 minutes and 15 seconds after launch.

The primary pressurization system of the propellant tanks for the S-IVB restart uses a helium heater. In this system, nine helium storage spheres in the liquid hydrogen tank contain gaseous helium charged to about 3,000 psi. This helium is passed through the heater which heats and expands the gas before it enters the propellant tanks. The heater operates on hydrogen and oxygen gas from the main propellant tanks.

The backup system consists of five ambient helium spheres mounted on the stage thrust structure. This system, controlled by the fuel re-pressurization control module, can repressurize the tanks in case the primary system fails. The restart will use the primary system. If that system fails, the backup system will be used.

Differences in Launch Vehicles for Apollo 10 and Apollo 11

The greatest difference between the Saturn V launch vehicle for Apollo 10 and the one for Apollo 11 is in the number of instrumentation measurements planned for the flight. Apollo 11 will be flying the operational configuration of instrumentation. Most research and development instrumentation has been removed, reducing the total number of measurements from 2,342 on Apollo 10 to 1,348 on Apollo 11. Measurements on Apollo 10, with Apollo 11 measurements in parentheses, were: S-IC, 672 (330); S-II, 980 (514); S-IVB, 386 (283); and IU, 298 (221).

The center engine of the S-II will be cut off early, as was done during the Apollo 10 flight, to eliminate the longitudinal oscillations reported by astronauts on the Apollo 9 mission. Cutting off the engine early on Apollo 10 was the simplest and quickest method of solving the problem.

APOLLO 11 CREW

Life Support Equipment - Space Suits

Apollo 11 crewmen will wear two versions of the Apollo space suit: an intravehicular pressure garment assembly worn by the command module pilot and the extravehicular pressure garment assembly worn by the commander and the lunar module pilot. Both versions are basically identical except that the extravehicular version has an integral thermal/meteoroid garment over the basic suit.

From the skin out, the basic pressure garment consists of a nomex comfort layer, a neoprene-coated nylon pressure bladder and a nylon restraint layer. The outer layers of the intravehicular suit are, from the inside out, nomex and two layers of Teflon-coated Beta cloth. The extravehicular integral thermal/meteoroid cover consists of a liner of two layers of neoprene-coated nylon, seven layers of Beta/Kapton spacer laminate, and an outer layer of Teflon-coated Beta fabric.

The extravehicular suit, together with a liquid cooling garment, portable life support system (PLSS), oxygen purge system, lunar extravehicular visor assembly and other components make up the extravehicular mobility unit (EMU). The EMU provides an extravehicular crewman with life support for a four-hour mission outside the lunar module without replenishing expendables. EMU total weight is 183 pounds. The intravehicular suit weighs 35.6 pounds.

Liquid cooling garment--A knitted nylon-spandex garment with a network of plastic tubing through which cooling water from the PLSS is circulated. It is worn next to the skin and replaces the constant wear-garment during EVA only.

Portable life support system--A backpack supplying oxygen at 3.9 psi and cooling water to the liquid cooling garment. Return oxygen is cleansed of solid and gas contaminants by a lithium hydroxide canister. The PLSS includes communications and telemetry equipment, displays and controls, and a main power supply. The PLSS is covered by a thermal insulation jacket. (Two stowed in LM).

Oxygen purge system--Mounted atop the PLSS, the oxygen purge system provides a contingency 30-minute supply of gaseous oxygen in two two-pound bottles pressurized to 5,880 psia. The system may also be worn separately on the front of the pressure garment assembly torso. It serves as a mount for the VHF antenna for the PLSS. (Two stowed in LM).

Lunar extravehicular visor assembly -- A polycarbonate shell and two visors with thermal control and optical coatings on them. The EVA visor is attached over the pressure helmet to provide impact, micrometeoroid, thermal and ultraviolet infrared light protection to the EVA crewman.

Extravehicular gloves--Built of an outer shell of Chromel-R fabric and thermal insulation to provide protection when handling extremely hot and cold objects. The finger tips are made of silicone rubber to provide the crewman more sensitivity.

A one-piece constant-wear garment, similar to "long johns", is worn as an undergarment for the space suit in intravehicular operations and for the inflight coveralls. The garment is porous-knit cotton with a waist-to-neck zipper for donning. Biomedical harness attach points are provided.

During periods out of the space suits, crewmen will wear two-piece Teflon fabric inflight coveralls for warmth and for pocket stowage of personal items.

Communications carriers ("Snoopy hats") with redundant microphones and earphones are worn with the pressure helmet; a lightweight headset is worn with the inflight coveralls.

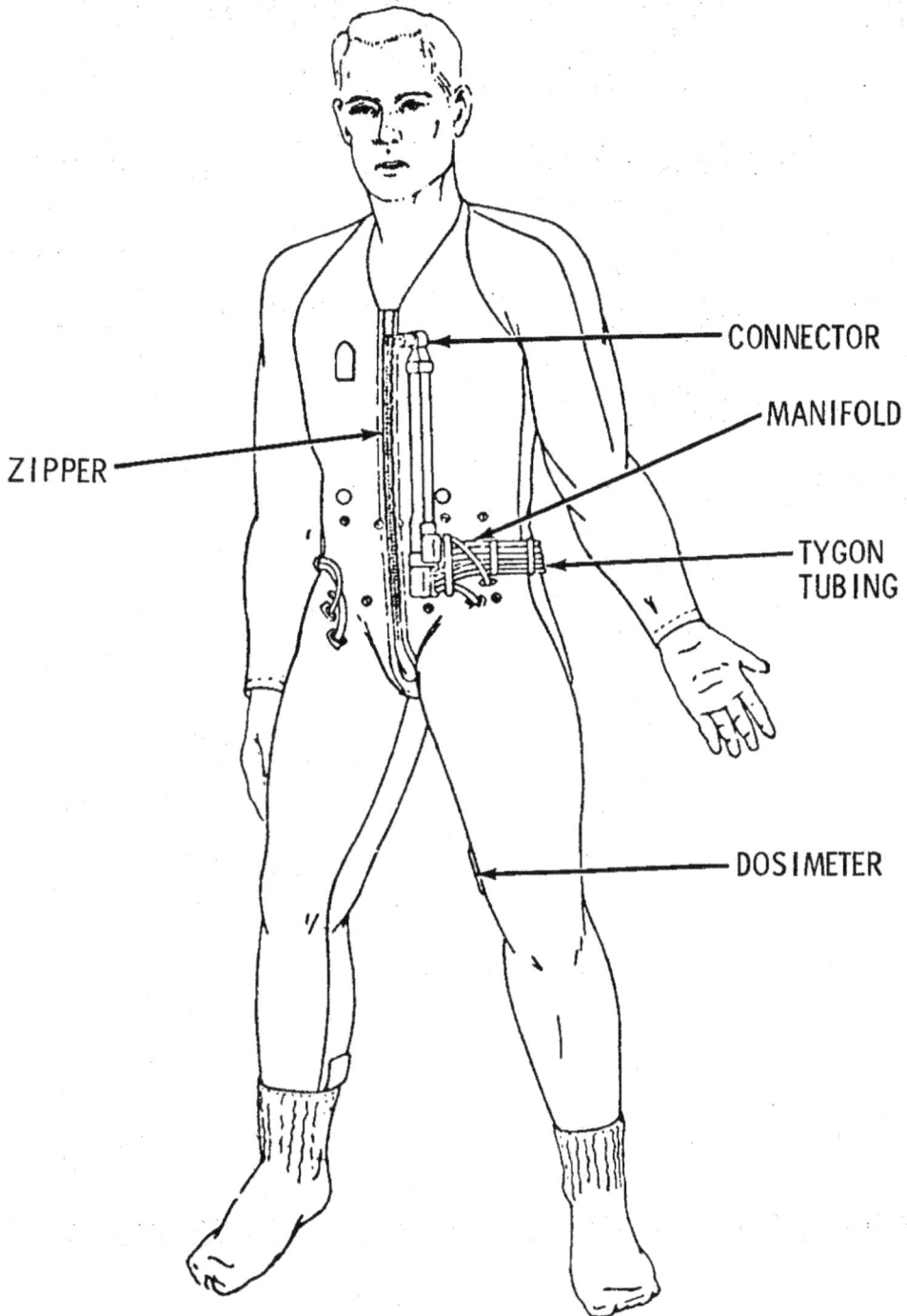

CONNECTOR

MANIFOLD

ZIPPER

TYGON TUBING

DOSIMETER

LIQUID COOLING GARMENT
-more-

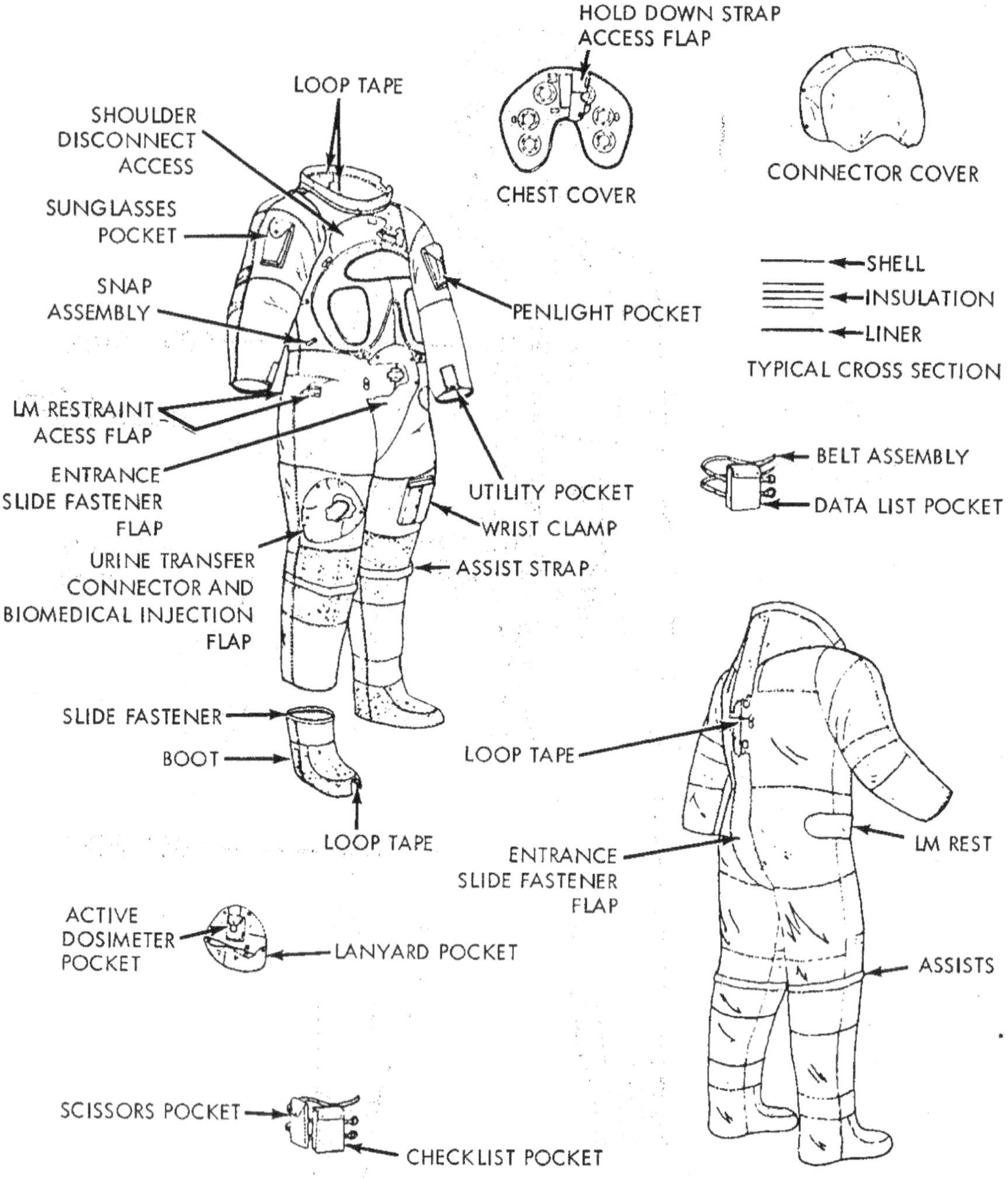

HOLD DOWN STRAP
ACCESS FLAP

CONNECTOR COVER

CHEST COVER

LOOP TAPE

SHOULDER
DISCONNECT
ACCESS

SUNGLASSES
POCKET

SNAP
ASSEMBLY

PENLIGHT POCKET

←SHELL
←INSULATION
←LINER

TYPICAL CROSS SECTION

LM RESTRAINT
ACESS FLAP

ENTRANCE
SLIDE FASTENER
FLAP

URINE TRANSFER
CONNECTOR AND
BIOMEDICAL INJECTION
FLAP

UTILITY POCKET

WRIST CLAMP

ASSIST STRAP

BELT ASSEMBLY

DATA LIST POCKET

SLIDE FASTENER

BOOT

LOOP TAPE

LOOP TAPE

ENTRANCE
SLIDE FASTENER
FLAP

LM REST

ACTIVE
DOSIMETER
POCKET

LANYARD POCKET

ASSISTS

SCISSORS POCKET

CHECKLIST POCKET

INTEGRATED THERMAL MICROMETEROID GARMENT

-more-

EXTRAVEHICULAR MOBILITY UNIT

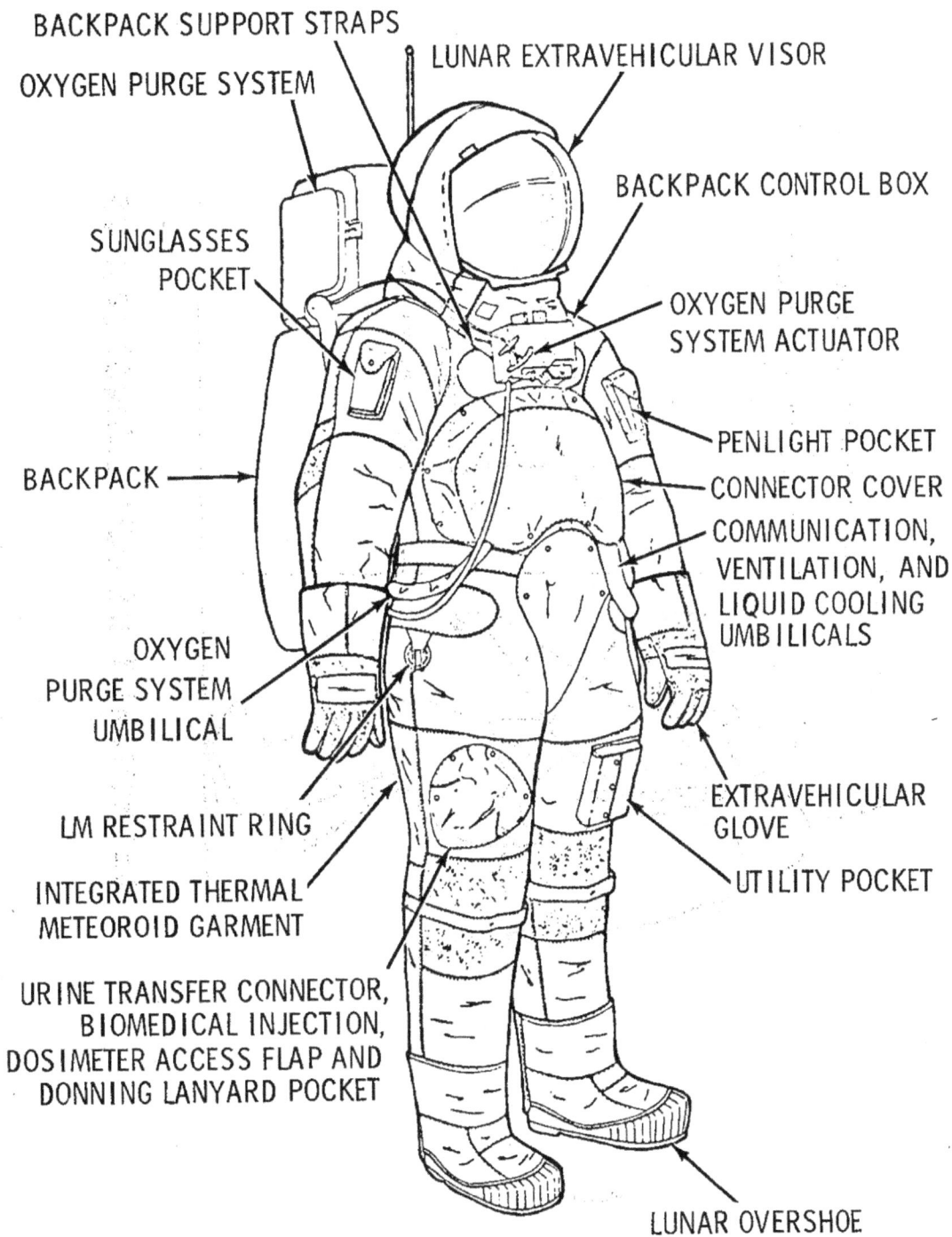

BACKPACK SUPPORT STRAPS

OXYGEN PURGE SYSTEM

LUNAR EXTRAVEHICULAR VISOR

BACKPACK CONTROL BOX

SUNGLASSES POCKET

OXYGEN PURGE SYSTEM ACTUATOR

PENLIGHT POCKET

CONNECTOR COVER

BACKPACK

COMMUNICATION, VENTILATION, AND LIQUID COOLING UMBILICALS

OXYGEN PURGE SYSTEM UMBILICAL

LM RESTRAINT RING

INTEGRATED THERMAL METEOROID GARMENT

EXTRAVEHICULAR GLOVE

UTILITY POCKET

URINE TRANSFER CONNECTOR, BIOMEDICAL INJECTION, DOSIMETER ACCESS FLAP AND DONNING LANYARD POCKET

LUNAR OVERSHOE

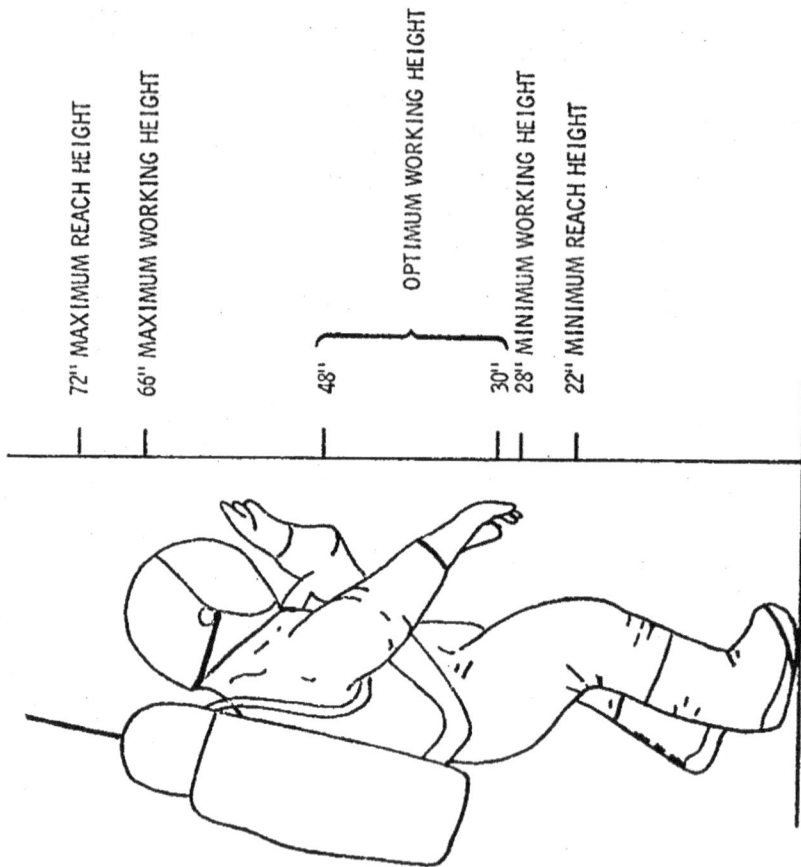

ASTRONAUT REACH CONSTRAINTS

72" MAXIMUM REACH HEIGHT

66" MAXIMUM WORKING HEIGHT

OPTIMUM WORKING HEIGHT

48"

30"

28" MINIMUM WORKING HEIGHT

22" MINIMUM REACH HEIGHT

APOLLO 11 CREW MENU

The Apollo 11 crew had a wide range of food items from which to select their daily mission space menu. More than 70 items comprise the food selection list of freeze-dried rehydratable, wet-pack and spoon-bowl foods.

Balanced meals for five days have been packed in man/day over-wraps, and items similar to those in the daily menus have been packed in a sort of snack pantry. The snack pantry permits the crew to locate easily a food item in a smorgasbord mode without having to "rob" a regular meal somewhere down deep in a storage box.

Water for drinking and rehydrating food is obtained from three sources in the command module—a dispenser for drinking water and two water spigots at the food preparation station, one supplying water at about 155 degrees F, the other at about 55 degrees F. The potable water dispenser squirts water continuously as long as the trigger is held down, and the food preparation spigots dispense water in one-ounce increments. Command module potable water is supplied from service module fuel cell byproduct water.

A continuous-feed hand water dispenser similar to the one in the command module is used aboard the lunar module for cold-water rehydration of food packets stowed aboard the LM.

After water has been injected into a food bag, it is kneaded for about three minutes. The bag neck is then cut off and the food squeezed into the crewman's mouth. After a meal, germicide pills attached to the outside of the food bags are placed in the bags to prevent fermentation and gas formation. The bags are then rolled and stowed in waste disposal compartments.

The day-by-day, meal-by-meal Apollo 11 menu for each crew-man as well as contents of the snack pantry are listed on the following pages:

APOLLO XI (ARMSTRONG)

MEAL	DAY 1*, 5	DAY 2	DAY 3	DAY 4
A	Peaches Bacon Squares (8) Strawberry Cubes (4) Grape Drink Orange Drink	Fruit Cocktail Sausage Patties** Cinn. Tstd. Bread Cubes (4) Cocoa Grapefruit Drink	Peaches Bacon Squares (8) Apricot Cereal Cubes (4) Grape Drink Orange Drink	Canadian Bacon and Applesauce Sugar Coated Corn Flakes Peanut Cubes (4) Cocoa Orange-Grapefruit Drink
B	Beef and Potatoes*** Butterscotch Pudding Brownies (4) Grape Punch	Frankfurters*** Applesauce Chocolate Pudding Orange-Grapefruit Drink	Cream of Chicken Soup Turkey and Gravy*** Cheese Cracker Cubes (6) Chocolate Cubes (6) Pineapple-Grapefruit Drink	Shrimp Cocktail Ham and Potatoes*** Fruit Cocktail Date Fruitcake (4) Grapefruit Drink
C	Salmon Salad Chicken and Rice** Sugar Cookie Cubes (6) Cocoa Pineapple-Grapefruit Drink	Spaghetti with Meat Sauce** Pork and Scalloped Potatoes** Pineapple Fruitcake (4) Grape Punch	Tuna Salad Chicken Stew** Butterscotch Pudding Cocoa Grapefruit Drink	Beef Stew** Coconut Cubes (4) Banana Pudding Grape Punch

-more-

*Day 1 consists of Meal B and C only
**Spoon-Bowl Package
***Wet-Pack Food

APOLLO XI (COLLINS)

MEAL	DAY 1*, 5	DAY 2	DAY 3	DAY 4
A	Peaches Bacon Squares (8) Strawberry Cubes (4) Grape Drink Orange Drink	Fruit Cocktail Sausage Patties** Cinn. Tstd. Bread Cubes (4) Cocoa Grapefruit Drink	Peaches Bacon Squares (8) Apricot Cereal Cubes (4) Grape Drink Orange Drink	Canadian Bacon and Applesauce Sugar Coated Corn Flakes Peanut Cubes (4) Cocoa Orange-Grapefruit Drink
B	Beef and Potatoes*** Butterscotch Pudding Brownies (4) Grape Punch	Frankfurters*** Applesauce Chocolate Pudding Orange-Grapefruit Drink	Cream of Chicken Soup Turkey and Gravy*** Cheese Cracker Cubes (6) Chocolate Cubes (4) Pineapple-Grapefruit Drink	Shrimp Cocktail Ham and Potatoes*** Fruit Cocktail Date Fruitcake (4) Grapefruit Drink
C	Salmon Salad Chicken and Rice** Sugar Cookie Cubes (6) Cocoa Pineapple-Grapefruit Drink	Potato Soup Pork and Scalloped Potatoes*** Pineapple Fruitcake (4) Grape Punch	Tuna Salad Chicken Stew** Butterscotch Pudding Cocoa Grapefruit Drink	Beef Stew** Coconut Cubes (4) Banana Pudding Grape Punch

*Day 1 consists of Meal B and C only
**Spoon-Bowl Package
***Wet-Pack Food

-more-

APOLLO XI (ALDRIN)

MEAL	DAY 1*, 5	DAY 2	DAY 3	DAY 4
A	Peaches Bacon Squares (8) Strawberry Cubes (4) Grape Drink Orange Drink	Fruit Cocktail Sausage Patties** Cinn. Tstd. Bread Cubes (4) Cocoa Grapefruit Drink	Peaches Bacon Squares (8) Apricot Cereal Cubes (4) Grape Drink Orange Drink	Canadian Bacon and Applesauce Sugar Coated Corn Flakes Peanut Cubes (4) Cocoa Orange-Grapefruit Drink
B	Beef and Potatoes*** Butterscotch Pudding Brownies (4) Grape Punch	Frankfurters*** Applesauce Chocolate Pudding Orange-Grapefruit Drink	Cream of Chicken Soup Turkey and Gravy*** Cheese Cracker Cubes (5) Chocolate Cubes (6) Pineapple-Grapefruit Drink	Shrimp Cocktail Ham and Potatoes*** Fruit Cocktail Date Fruitcake (4) Grapefruit Drink
C	Salmon Salad Chicken and Rice** Sugar Cookie Cubes (4) Cocoa Pineapple-Grapefruit Drink	Chicken Salad Chicken and Gravy Beef Sandwiches (6) Pineapple Fruitcake (4) Grape Punch	Tuna Salad Chicken Stew** Butterscotch Pudding Cocoa Grapefruit Drink	Pork and Scalloped Potatoes** Coconut Cubes (4) Banana Pudding Grape Punch

-more-

*Day 1 consists of Meal B and C only
**Spoon-Bowl Package
***Wet-Pack Food

ACCESSORIES	Unit
Chewing gum	15
Wet skin cleaning towels	30
Oral Hygiene Kit	1
3 toothbrushes	
1 edible toothpaste	
1 dental floss	
Contingency Feeding System	1
3 food restrainer pouches	
3 beverage packages	
1 valve adapter (pontube)	
Spoons	3

-more-

Snack Pantry

Breakfast	Units
Peaches	6
Fruit Cocktail	6
Canadian Bacon and Applesauce	3
Bacon Squares (8)	12
Sausage Patties*	3
Sugar Coated Corn Flakes	6
Strawberry Cubes (4)	3
Cinn. Tstd. Bread Cubes (4)	6
Apricot Cereal Cubes (4)	3
Peanut Cubes (4)	3
	51

Salads/Meats	
Salmon Salad	3
Tuna Salad	3
Cream of Chicken Soup	6
Shrimp Cocktail	6
Spaghetti and Meat Sauce*	6
Beef Pot Roast	3
Beef and Vegetables	3
Chicken and Rice*	6
Chicken Stew*	3
Beef Stew*	3
Pork and Scalloped Potatoes*	6
Ham and Potatoes (Wet)	3
Turkey and Gravy (Wet)	6
	57

*Spoon-Bowl Package

-more-

Snack Pantry

Rehydratable Desserts	Units
Banana Pudding	6
Butterscotch Pudding	6
Applesauce	6
Chocolate Pudding	6
	24

Beverages	
Orange Drink	6
Orange-Grapefruit Drink	3
Pineapple-Grapefruit Drink	3
Grapefruit Drink	3
Grape Drink	6
Grape Punch	3
Cocoa	6
Coffee (B)	15
Coffee (S)	15
Coffee (C and S)	15
	75

-more-

Snack Pantry

Dried Fruits	Units	Stow
Apricots	6	1
Peaches	6	1
Pears	6	1

Sandwich Spread

	Units	Stow
Ham Salad (5 oz.)	1	1
Tuna Salad (5 oz.)	1	1
Chicken Salad (5 oz.)	1	1
Cheddar Cheese (2 oz.)	3	1

Bread

	Units	Stow
Rye	6	6
White	6	6

-more-

Snack Pantry

Bites	Units
Cheese Cracker Cubes (6)	6
BBQ Beef Bits (4)	6
Chocolate Cubes (4)	6
Brownies (4)	6
Date Fruitcake (4)	6
Pineapple Fruitcake (4)	6
Jellied Fruit Candy (4)	6
Caramel Candy (4)	6

LM-5 Food

Meal A.   Bacon Squares(8)

                Peaches

                Sugar Cookie Cubes (6)

                Coffee

                Pineapple-Grapefruit drink

Meal B.   Beef stew

                Cream of Chicken Soup

                Date Fruit Cake (4)

                Grape Punch

                Orange Drink

	Units
Extra Beverage	8
Dried Fruit	4
Candy Bar	4
Bread	2
Ham Salad Spread (tube food)	1
Turkey and Gravy	2
Spoons	2

-more-

Personal Hygiene

Crew personal hygiene equipment aboard Apollo 11 includes body cleanliness items, the waste management system and one medical kit.

Packaged with the food are a toothbrush and a two-ounce tube of toothpaste for each crewman. Each man-meal package contains a 3.5-by-four-inch wet-wipe cleansing towel. Additionally, three packages of 12-by-12-inch dry towels are stowed beneath the command module pilot's couch. Each package contains seven towels. Also stowed under the command module pilot's couch are seven tissue dispensers containing 53 three-ply tissues each.

Solid body wastes are collected in Gemini-type plastic defecation bags which contain a germicide to prevent bacteria and gas formation. The bags are sealed after use and stowed in empty food containers for post-flight analysis.

Urine collection devices are provided for use while wearing either the pressure suit or the inflight coveralls. The urine is dumped overboard through the spacecraft urine dump valve in the CM and stored in the LM.

Medical Kit

The 5x5x8-inch medical accessory kit is stowed in a compartment on the spacecraft right side wall beside the lunar module pilot couch. The medical kit contains three motion sickness injectors, three pain suppression injectors, one two-ounce bottle first aid ointment, two one-ounce bottle eye drops, three nasal sprays, two compress bandages, 12 adhesive bandages, one oral thermometer and four spare crew biomedical harnesses. Pills in the medical kit are 60 antibiotic, 12 nausea, 12 stimulant, 18 pain killer, 60 decongestant, 24 diarrhea, 72 aspirin and 21 sleeping. Additionally, a small medical kit containing four stimulant, eight diarrhea, two sleeping and four pain killer pills, 12 aspirin, one bottle eye drops and two compress bandages is stowed in the lunar module flight data file compartment.

Survival Gear

The survival kit is stowed in two rucksacks in the right-hand forward equipment bay above the lunar module pilot.

Contents of rucksack No. 1 are: two combination survival lights, one desalter kit, three pair sunglasses, one radio beacon, one spare radio beacon battery and spacecraft connector cable, one knife in sheath, three water containers and two containers of Sun lotion.

RUCKSACK A

RUCKSACK B

DYE MARKER

3-MAN LIFE RAFT WITH SUN BONNETS

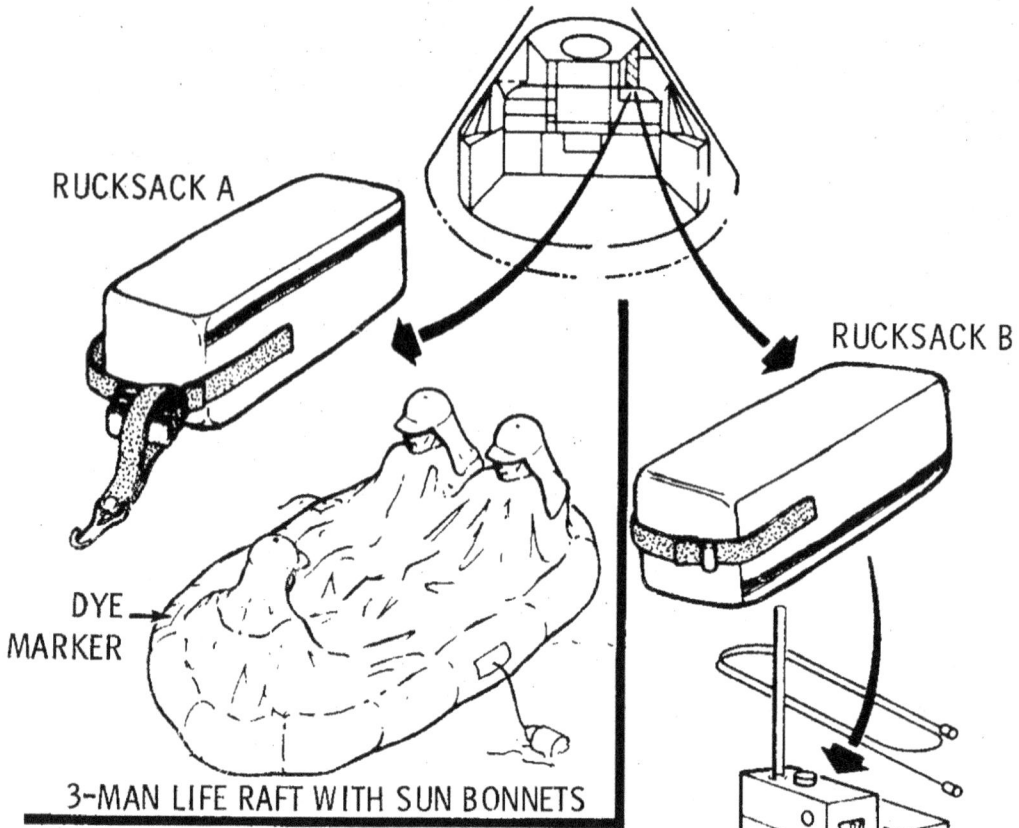

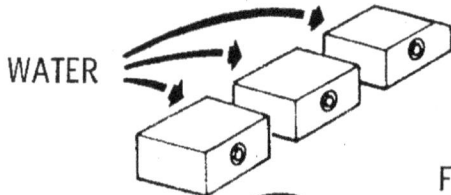

BEACON TRANSCEIVER, BATTERY AND CABLE

WATER

FIRST AID KIT

TABLETS(16)

SURVIVAL GLASSES (3)

DESALTING KITS (2)

SURVIVAL KNIFE

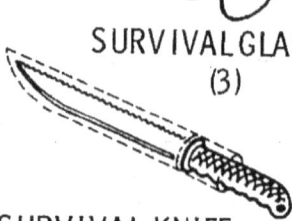

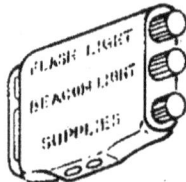

SURVIVAL LIGHTS

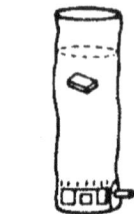

-more-

Rucksack No. 2: one three-man life raft with CO_2 inflater, one sea anchor, two sea dye markers, three sun-bonnets, one mooring lanyard, three manlines, and two attach brackets.

The survival kit is designed to provide a 48-hour postlanding (water or land) survival capability for three crewmen between 40 degrees North and South latitudes.

Biomedical Inflight Monitoring

The Apollo 11 crew biomedical telemetry data received by the Manned Space Flight Network will be relayed for instantaneous display at Mission Control Center where heart rate and breathing rate data will be displayed on the flight surgeon's console. Heart rate and respiration rate average, range and deviation are computed and displayed on digital TV screens.

In addition, the instantaneous heart rate, real-time and delayed EKG and respiration are recorded on strip charts for each man.

Biomedical telemetry will be simultaneous from all crewmen while in the CSM, but selectable by a manual onboard switch in the LM.

Biomedical data observed by the flight surgeon and his team in the Life Support Systems Staff Support Room will be correlated with spacecraft and space suit environmental data displays.

Blood pressures are no longer telemetered as they were in the Mercury and Gemini programs. Oral temperature, however, can be measured onboard for diagnostic purposes and voiced down by the crew in case of inflight illness.

Training

The crewmen of Apollo 11 have spent more than five hours of formal crew training for each hour of the lunar-orbit mission's eight-day duration. More than 1,000 hours of training were in the Apollo 11 crew training syllabus over and above the normal preparations for the mission--technical briefings and reviews, pilot meetings and study.

The Apollo 11 crewmen also took part in spacecraft manufacturing checkouts at the North American Rockwell plant in Downey, Calif., at Grumman Aircraft Engineering Corp., Bethpage, N.Y., and in prelaunch testing at NASA Kennedy Space Center. Taking part in factory and launch area testing has provided the crew with thorough operational knowledge of the complex vehicle.

Highlights of specialized Apollo 11 crew training topics are:

* Detailed series of briefings on spacecraft systems, operation and modifications.

* Saturn launch vehicle briefings on countdown, range safety, flight dynamics, failure modes and abort conditions. The launch vehicle briefings were updated periodcally.

* Apollo Guidance and Navigation system briefings at the Massachusetts Institute of Technology Instrumentation Laboratory.

* Briefings and continuous training on mission photographic objectives and use of camera equipment.

* Extensive pilot participation in reviews of all flight procedures for normal as well as emergency situations.

* Stowage reviews and practice in training sessions in the spacecraft, mockups and command module simulators allowed the crewmen to evaluate spacecraft stowage of crew-associated equipment.

* More than 400 hours of training per man in command module and lunar module simulators at MSC and KSC, including closed-loop simulations with flight controllers in the Mission Control Center. Other Apollo simulators at various locations were used extensively for specialized crew training.

* Entry corridor deceleration profiles at lunar-return conditions in the MSC Flight Acceleration Facility manned centrifuge.

• Lunar surface briefings and 1-g walk-throughs of lunar surface EVA operations covering lunar geology and microbiology and deployment of experiments in the Early Apollo Surface Experiment Package (EASEP). Training in lunar surface EVA included practice sessions with lunar surface sample gathering tools and return containers, cameras, the erectable S-band antenna and the modular equipment stowage assembly (MESA) housed in the LM descent stage.

• Proficiency flights in the lunar landing training vehicle (LLTV) for the commander.

* Zero-g aircraft flights using command module and lunar module mockups for EVA and pressure suit doffing/donning practice and training.

* Underwater zero-g training in the MSC Water Immersion Facility using spacecraft mockups to further familiarize crew with all aspects of CSM-LM docking tunnel intravehicular transfer and EVA in pressurized suits.

* Water egress training conducted in indoor tanks as well as in the Gulf of Mexico, included uprighting from the Stable II position (apex down) to the Stable I position (apex up), egress onto rafts and helicopter pickup.

* Launch pad egress training from mockups and from the actual spacecraft on the launch pad for possible emergencies such as fire, contaminants and power failures.

* The training covered use of Apollo spacecraft fire suppression equipment in the cockpit.

* Planetarium reviews at Morehead Planetarium, Chapel Hill, N.C., and at Griffith Planetarium, Los Angeles, Calif., of the celestial sphere with special emphasis on the 37 navigational stars used by the Apollo guidance computer.

NATIONAL AERONAUTICS AND SPACE ADMINISTRATION

WASHINGTON, D. C. 20546

BIOGRAPHICAL DATA

NAME: Neil A. Armstrong (Mr.)
NASA Astronaut, Commander, Apollo 11

BIRTHPLACE AND DATE: Born in Wapakoneta, Ohio, on August 5, 1930; he is the son of Mr. and Mrs. Stephen Armstrong of Wapakoneta.

PHYSICAL DESCRIPTION: Blond hair; blue eyes; height: 5 feet 11 inches; weight: 165 pounds.

EDUCATION: Attended secondary school in Wapakoneta, Ohio; received a Bachelor of Science degree in Aeronautical Engineering from Purdue University in 1955. Graduate School - University of Southern California.

MARITAL STATUS: Married to the former Janet Shearon of Evanston, Illinois, who is the daughter of Mrs. Louise Shearon of Pasadena, California.

CHILDREN: Eric, June 30, 1957; Mark, April 8, 1963.

OTHER ACTIVITIES: His hobbies include soaring (for which he is a Federation Aeronautique Internationale gold badge holder).

ORGANIZATIONS: Associate Fellow of the Society of Experimental Test Pilots; associate fellow of the American Institute of Aeronautics and Astronautics; and member of the Soaring Society of America.

SPECIAL HONORS: Recipient of the 1962 Institute of Aerospace Sciences Octave Chanute Award; the 1966 AIAA Astronautics Award; the NASA Exceptional Service Medal; and the 1962 John J. Montgomery Award.

EXPERIENCE: Armstrong was a naval aviator from 1949 to 1952 and flew 78 combat missions during the Korean action.

He joined NASA's Lewis Research Center in 1955 (then NACA Lewis Flight Propulsion Laboratory) and later transferred to the NASA High Speed Flight Station (now Flight Research Center) at Edwards Air Force Base, California, as an aeronautical research pilot for NACA and NASA. In this capacity, he performed as an X-15 project pilot, flying that aircraft to over 200,000 feet and approximately 4,000 miles per hour.

Other flight test work included piloting the X-1 rocket
airplane, the F-100, F-101, F-102, F-104, F5D, B-47, the
paraglider, and others.

As pilot of the B-29 "drop" aircraft, he participated in
the launches of over 100 rocket airplane flights.

He has logged more than 4,000 hours flying time.

CURRENT ASSIGNMENT: Mr. Armstrong was selected as an astronaut
by NASA in September 1962. He served as backup command
pilot for the Gemini 5 flight.

As command pilot for the Gemini 8 mission, which was launched
on March 16, 1966, he performed the first successful docking of two vehicles in space. The flight, originally
scheduled to last three days, was terminated early due to
a malfunctioning OAMS thruster; but the crew demonstrated
exceptional piloting skill in overcoming this problem
and bringing the spacecraft to a safe landing.

He subsequently served as backup command pilot for the
Gemini 11 mission and is currently assigned as the
commander for the Apollo 11 mission, and will probably
be the first human to set foot on the Moon.

As a civil servant, Armstrong, a GS-16 Step 7, earns
$30,054 per annum

NATIONAL AERONAUTICS AND SPACE ADMINISTRATION

WASHINGTON, D. C. 20546

BIOGRAPHICAL DATA

NAME: Michael Collins (Lieutenant Colonel, USAF)
       NASA Astronaut, Command Module Pilot, Apollo 11

BIRTHPLACE AND DATE: Born in Rome, Italy, on October 31, 1930.
       His mother, Mrs. James L. Collins, resides in Washington,
       D.C.

PHYSICAL DESCRIPTION: Brown hair; brown eyes; height: 5 feet
       11 inches; weight: 165 pounds.

EDUCATION: Graduated from Saint Albans School in Washington,
       D.C.; received a Bachelor of Science degree from the
       United States Military Academy at West Point, New York,
       in 1952.

MARITAL STATUS: Married to the former Patricia M. Finnegan of
       Boston, Massachusetts.

CHILDREN: Kathleen, May 6, 1959; Ann S., October 31, 1961;
       Michael L., February 23, 1963.

OTHER ACTIVITIES: His hobbies include fishing and handball.

ORGANIZATIONS: Member of the Society of Experimental Test Pilots.

SPECIAL HONORS: Awarded the NASA Exceptional Service Medal, the
       Air Force Command Pilot Astronaut Wings, and the Air Force
       Distinguished Flying Cross.

EXPERIENCE: Collins, an Air Force Lt. Colonel, chose an Air Force
       career following graduation from West Point.

       He served as an experimental flight test officer at the Air
       Force Flight Test Center, Edwards Air Force Base, California,
       and, in that capacity, tested performance and stability and
       control characteristics of Air Force aircraft--primarily
       jet fighters.

       He has logged more than 4,000 hours flying time, including
       more than 3,200 hours in jet aircraft.

CURRENT ASSIGNMENT: Lt. Colonel Collins was one of the third
group of astronauts named by NASA in October 1963. He
has since served as backup pilot for the Gemini 7
mission.

As pilot on the 3-day 44-revolution Gemini 10 mission,
launched July 18, 1966, Collins shares with command
pilot John Young in the accomplishments of that record-
setting flight. These accomplishments include a successful rendezvous and docking with a separately launched
Agena target vehicle and, using the power of the Agena,
maneuvering the Gemini spacecraft into another orbit for
a rendezvous with a second, passive Agena. Collins' skill-
ful performance in completing two periods of extravehicular
activity, including his recovery of a micrometeorite
detection experiment from the passive Agena, added greatly
to our knowledge of manned space flight.

Gemini 10 attained an apogee of approximately 475 statute
miles and traveled a distance of 1,275,091 statute miles--
after which splashdown occurred in the West Atlantic
529 statute miles east of Cape Kennedy. The spacecraft
landed 2.6 miles from the USS GUADALCANAL and became the
second in the Gemini program to land within eye and camera
range of a prime recovery vessel.

He is currently assigned as command module pilot on the
Apollo 11 mission. The annual pay and allowances of an
Air Force lieutenant colonel with Collins' time in service
totals $17,147.36.

NATIONAL AERONAUTICS AND SPACE ADMINISTRATION

WASHINGTON, D. C. 20546

BIOGRAPHICAL DATA

NAME: Edwin E. Aldrin, Jr. (Colonel, USAF)
NASA Astronaut, Lunar Module Pilot, Apollo 11

BIRTHPLACE AND DATE: Born in Montclair, New Jersey, on January 20, 1930, and is the son of the late Marion Moon Aldrin and Colonel (USAF Retired) Edwin E. Aldrin, who resides in Brielle, New Jersey.

PHYSICAL DESCRIPTION: Blond hair; blue eyes; height: 5 feet 10 inches; weight: 165 pounds.

EDUCATION: Graduated from Montclair High School, Montclair, New Jersey; received a Bachelor of Science degree from the United States Military Academy at West Point, New York, in 1951 and a Doctor of Science degree in Astronautics from the Massachusetts Institute of Technology in 1963; recipient of an Honorary Doctorate of Science degree from Gustavus Adolphus College in 1967, Honorary degree from Clark University, Worchester, Mass.

MARITAL STATUS: Married to the former Joan A. Archer of Ho-Ho-Kus, New Jersey, whose parents, Mr. and Mrs. Michael Archer, are residents of that city.

CHILDREN: J. Michael, September 2, 1955; Janice R., August 16, 1957; Andrew J., June 17, 1958.

OTHER ACTIVITIES: He is a Scout Merit Badge Counsellor and an Elder and Trustee of the Webster Presbyterian Church. His hobbies include running, scuba diving, and high bar exercises.

ORGANIZATIONS: Associate Fellow of the American Institute of Aeronautics and Astronautics; member of the Society of Experimental Test Pilots, Sigma Gamma Tau (aeronautical engineering society), Tau Beta Pi (national engineering society), and Sigma Xi (national science research society); and a 32nd Degree Mason advanced through the Commandery and Shrine.

SPECIAL HONORS: Awarded the Distinguished Flying Cross with one
     Oak Leaf Cluster, the Air Medal with two Oak Leaf Clusters,
     the Air Force Commendation Medal, the NASA Exceptional
     Service Medal and Air Force Command Pilot Astronaut Wings,
     the NASA Group Achievement Award for Rendezvous Operations
     Planning Team, an Honorary Life Membership in the Inter-
     national Association of Machinists and Aerospace Workers,
     and an Honorary Membership in the Aerospace Medical
     Association.

EXPERIENCE: Aldrin, an Air Force Colonel, was graduated third
     in a class of 475 from the United States Military Academy
     at West Point in 1951 and subsequently received his wings
     at Bryan, Texas, in 1952.

     He flew 66 combat missions in F-86 aircraft while on duty
     in Korea with the 51st Fighter Interceptor Wing and was
     credited with destroying two MIG-15 aircraft. At Nellis
     Air Force Base, Nevada, he served as an aerial gunnery
     instructor and then attended the Squadron Officers' School
     at the Air University, Maxwell Air Force Base, Alabama.

     Following his assignment as Aide to the Dean of Faculty at
     the United States Air Force Academy, Aldrin Flew F-100
     aircraft as a flight commander with the 36th Tactical
     Fighter Wing at Bitburg, Germany. He attended MIT,
     receiving a doctorate after completing his thesis concern-
     ing guidance for manned orbital rendezvous, and was then
     assigned to the Gemini Target Office of the Air Force Space
     Systems Division, Los Angeles, Calfornia. He was later
     transferred to the USAF Field Office at the Manned Space-
     craft Center which was responsible for integrating DOD
     experiments into the NASA Gemini flights.

     He has logged approximately 3,500 hours flying time,
     including 2,853 hours in jet aircraft and 139 hours in
     helicopters. He has made several flights in the lunar
     landing research vehicle.

CURRENT ASSIGNMENT: Colonel Aldrin was one of the third group
     of astronauts named by NASA in October 1963. He has since
     served as backup pilot for the Gemini 9 mission and prime
     pilot for the Gemini 12 mission.

-more-

On November 11, 1966, he and command pilot James Lovell
were launched into space in the Gemini 12 spacecraft on
a 4-day 59-revolution flight which brought the Gemini
Program to a successful close. Aldrin established a new
record for extravehicular activity (EVA) by accruing
slightly more than $5\frac{1}{2}$ hours outside the spacecraft. During
the umbilical EVA, he attached a tether to the Agena;
retrieved a micro-meteorite experiment package from the
spacecraft; and evaluated the use of body restraints
specially designed for completing work tasks outside the
spacecraft. He completed numerous photographic experiments
and obtained the first pictures taken from space of an
eclipse of the sun.

Other major accomplishments of the 94-hour 35-minute flight
included a third-revolution rendezvous with the previously
launched Agena, using for the first time backup onboard
computations due to a radar failure, and a fully automatic
controlled reentry of a spacecraft. Gemini 12 splashed
down in the Atlantic within $2\frac{1}{2}$ miles of the prime recovery
ship USS WASP.

Aldrin is currently assigned as lunar module pilot for the
Apollo 11 flight. The annual pay and allowances of an Air
Force colonel with Aldrin's time in service total $18,622.56.

-more-

EARLY APOLLO SCIENTIFIC EXPERIMENTS PACKAGE (EASEP)

The Apollo 11 scientific experiments for deployment on the lunar surface near the touchdown point of the lunar module are stowed in the LM's scientific equipment bay at the left rear quadrant of the descent stage looking forward.

The Early Apollo Scientific Experiments Package (EASEP) will be carried only on Apollo 11; subsequent Apollo lunar landing missions will carry the more comprehensive Apollo Lunar Surface Experiment Package (ALSEP).

EASEP consists of two basic experiments: the passive seismic experiments package (PSEP) and the laser ranging retro-reflector (LRRR). Both experiments are independent, self-contained packages that weigh a total of about 170 pounds and occupy 12 cubic feet of space.

PSEP uses three long-period seismometers and one short-period vertical seismometer for measuring meteoroid impacts and moonquakes. Such data will be useful in determining the interior structure of the Moon; for example, does the Moon have a core and mantle like Earth.

The seismic experiment package has four basic subsystems: structure/thermal subsystem for shock, vibration and thermal protection; electrical power subsystem generates 34 to 46 watts by solar panel array; data subsystem receives and decodes MSFN uplink commands and downlinks experiment data, handles power switching tasks; passive seismic experiment subsystem measures lunar seismic activity with long-period and short-period seismometers which detect inertial mass displacement.

The laser ranging retro-reflector experiment is a retro-reflector array with a folding support structure for aiming and aligning the array toward Earth. The array is built of cubes of fused silica. Laser ranging beams from Earth will be reflected back to their point of origin for precise measurement of Earth-Moon distances, motion of the Moon's center of mass, lunar radius and Earth geophysical information.

Earth stations which will beam lasers to the LRRR include the McDonald Observatory at Ft. Davis, Tex.; Lick Observatory, Mt. Hamilton, Calif.; and the Catalina Station of the University of Arizona. Scientists in other countries also plan to bounce laser beams off the LRRR.

Principal investigators for these experiments are Dr. C. O. Alley, University of Maryland (Laser Ranging Retro Feflector) and Dr. Garry Latham, Lamont Geological Observatory (Passive Seismic Experiments Package).

-more-

EASEP/LM INTERFACE

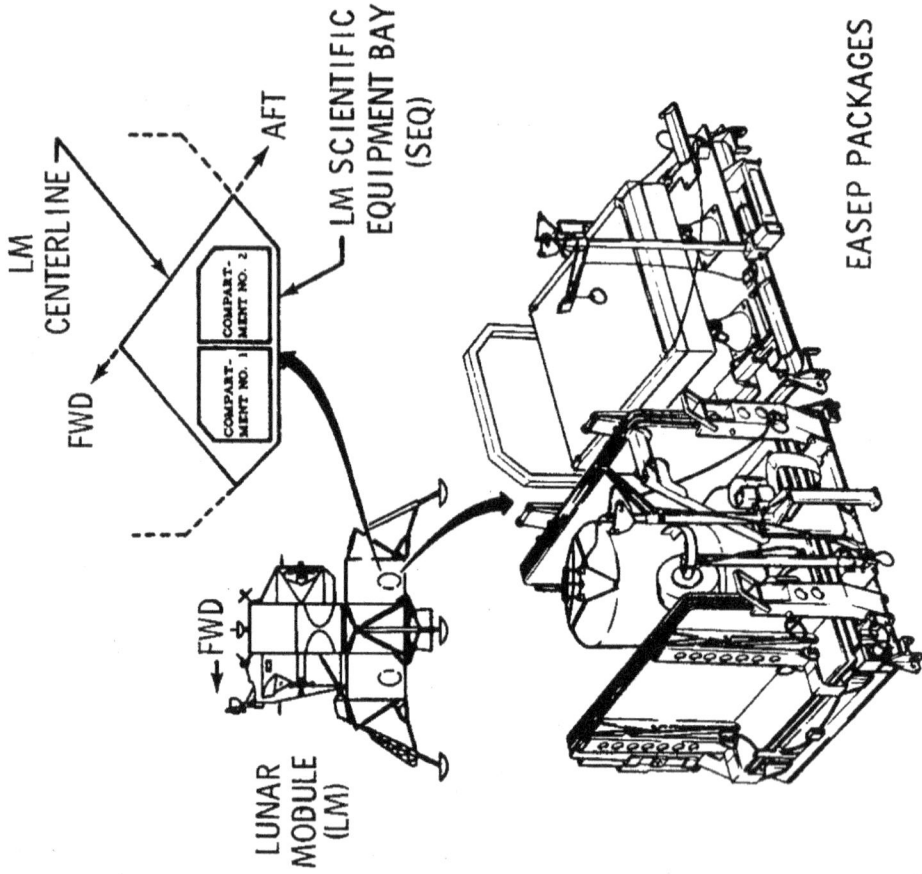

LM CENTERLINE

AFT

LM SCIENTIFIC
EQUIPMENT BAY
(SEQ)

FWD

COMPART-
MENT NO. 1

COMPART-
MENT NO. 2

FWD

LUNAR
MODULE (LM)

EASEP PACKAGES

EASEP DEPLOYMENT ZONES

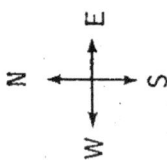

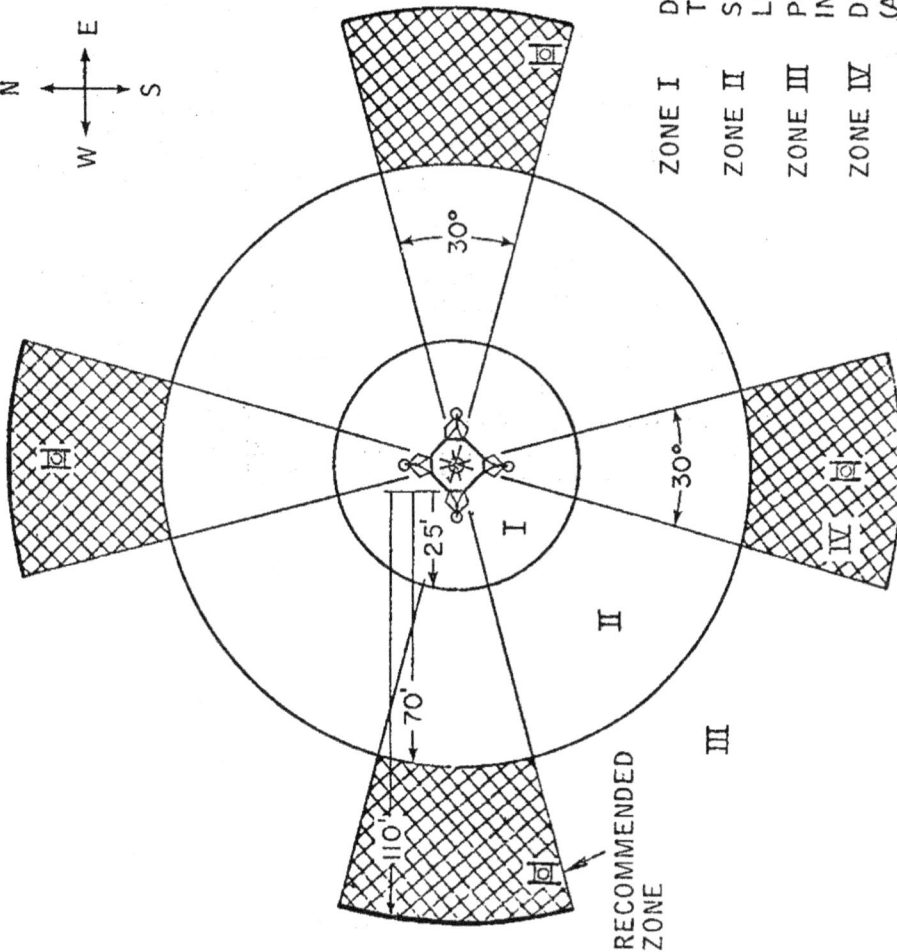

ZONE I   DEPLOYMENT OF EASEP RESTRICTED BY THERMAL RADIATION FROM THE LM

ZONE II   SIGNIFICANT AERODYNAMIC HEATING FROM LM ASCENT ENGINE PLUME

ZONE III   POSSIBLE CONTAMINATION BY KAPTON AND INCONEL DEBRIS

ZONE IV   DESIRABLE DEPLOYMENT ZONES (AVOID LM SHADOW)

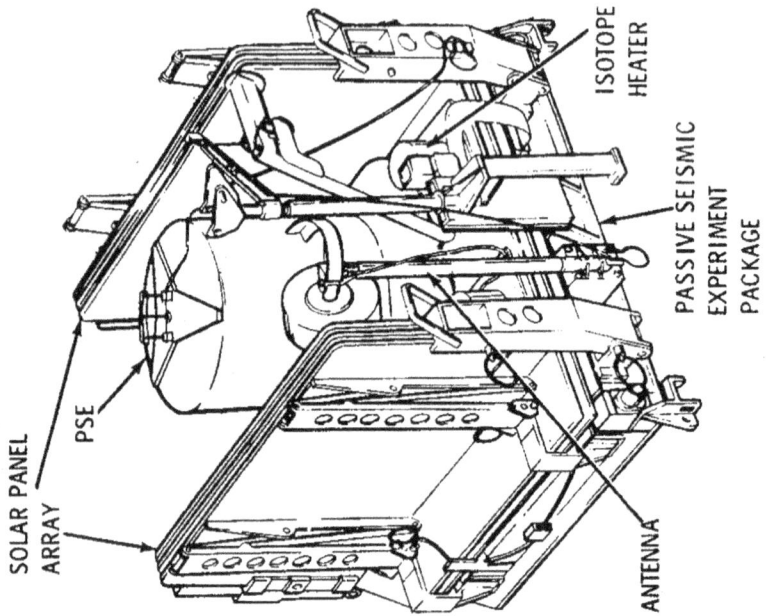

PSEP STOWED CONFIGURATION

SOLAR PANEL ARRAY

PSE

ISOTOPE HEATER

PASSIVE SEISMIC EXPERIMENT PACKAGE

ANTENNA

PSEP DEPLOYED CONFIGURATION

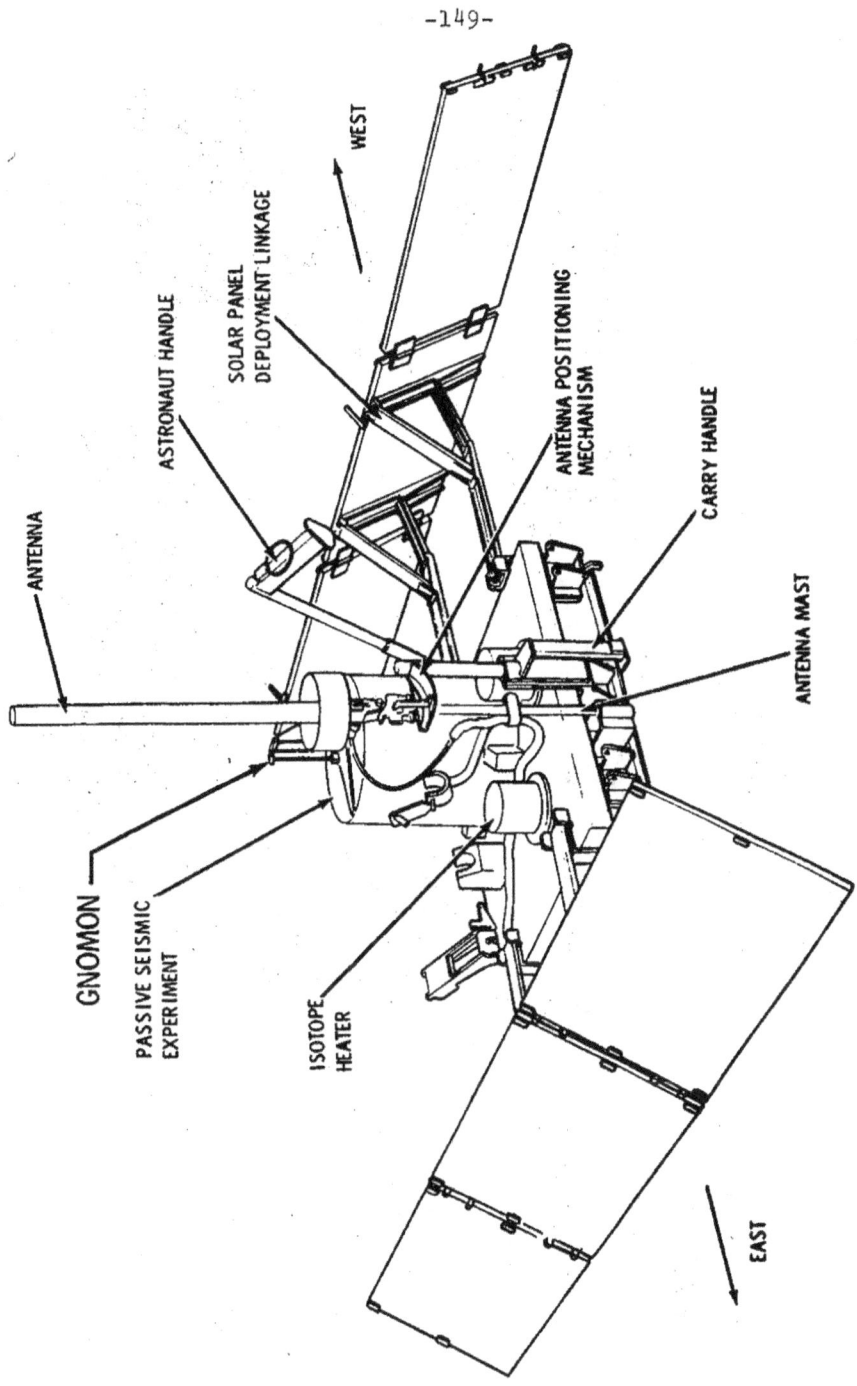

LASER RANGING RETRO-REFLECTOR EXPERIMENT

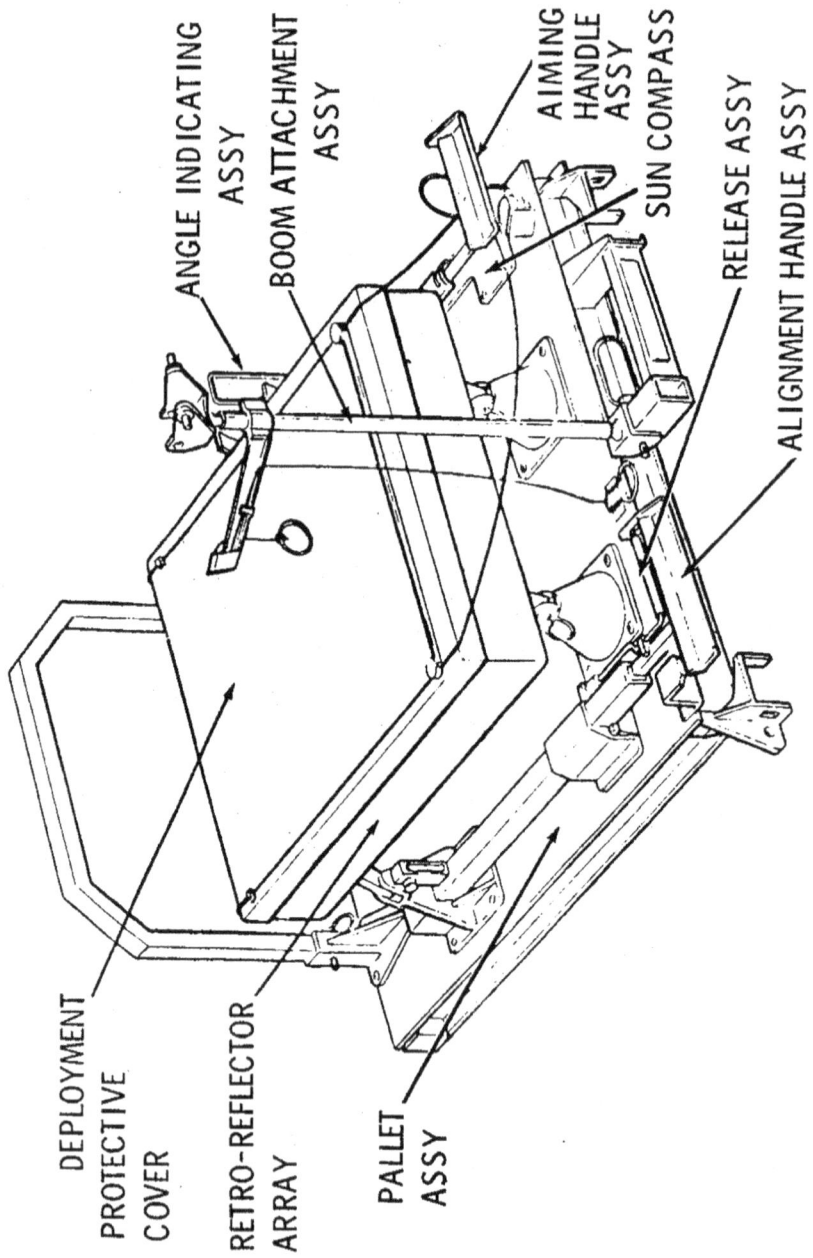

ANGLE INDICATING ASSY

BOOM ATTACHMENT ASSY

AIMING HANDLE ASSY

SUN COMPASS

RELEASE ASSY

ALIGNMENT HANDLE ASSY

DEPLOYMENT PROTECTIVE COVER

RETRO-REFLECTOR ARRAY

PALLET ASSY

LRRR DETAILS

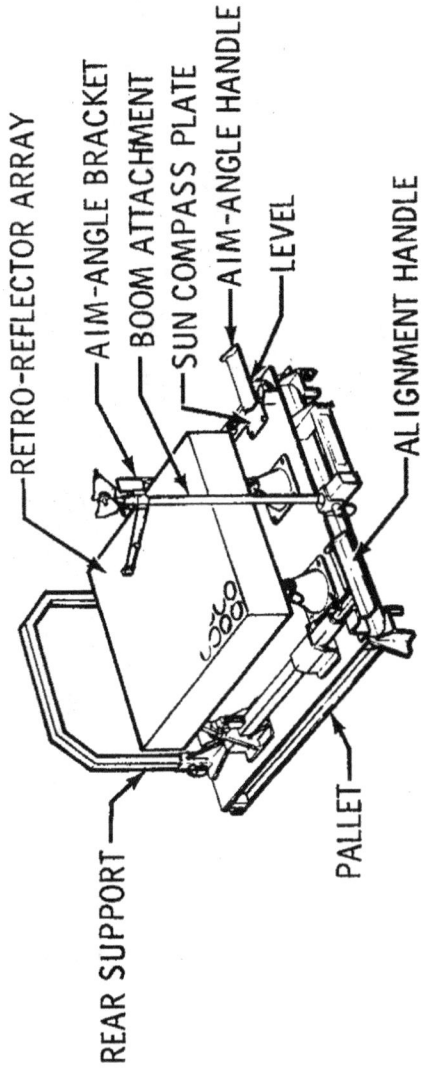

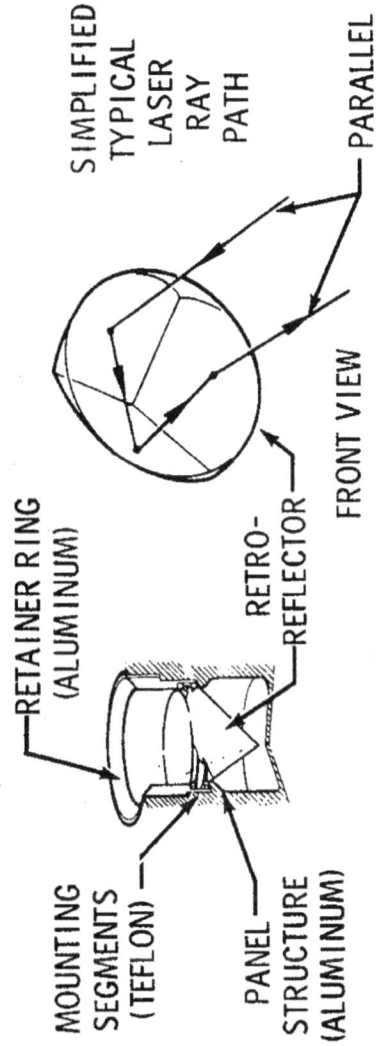

SOLAR WIND EXPERIMENT

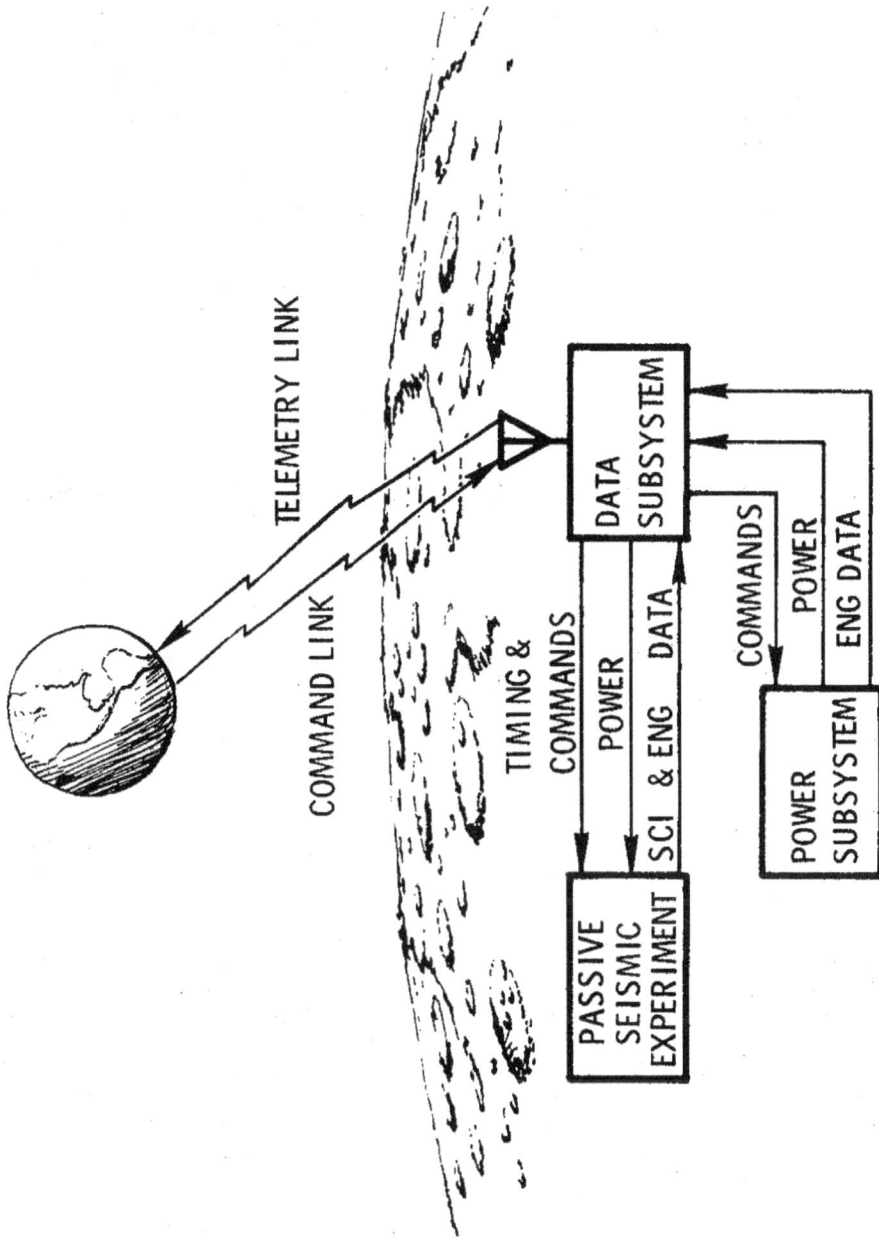

COMMAND AND TELEMETRY LINKS

APOLLO LUNAR RADIOISOTOPIC HEATER (ALRH)

An isotopic heater system built into the passive seismometer experiment package which Apollo 11 astronauts will leave on the Moon will protect the seismic recorder during frigid lunar nights.

The Apollo Lunar Radioisotopic Heater (ALRH), developed by the Atomic Energy Commission, will be the first major use of nuclear energy in a manned space flight mission. Each of the two heaters is fueled with about 1.2 ounces of plutonium 238. Heat is given off as the well shielded radioactive material decays.

During the lunar day, the seismic device will send back to Earth data on any lunar seismic activity or "Moonquakes." During the 340-hour lunar night, when temperatures drop as low as 279 degrees below zero F., the 15-watt heaters will keep the seismometer at a minimum of -65 degrees below zero F. Exposure to lowr temperatures would damage the device.

Power for the seismic experiment, which operates only during the day, is from two solar panels.

The heaters are three inches in diameter, three inches long, and weigh two pounds and two ounces each including multiple layers of shielding and protective materials. The complete seismometer package weighs 100 pounds.

They are mounted into the seismic package before launch. The entire unit will be carried in the lunar module scientific equipment bay and after landing on the Moon will be deployed by an astronaut a short distance from the lunar vehicle. There is no handling risk to the astronaut.

They are mounted into the seismic package before launch. The entire unit will be carried in the lunar module scientific by an astronaut a short distance from the lunar vehicle. There is no handling risk to the astronaut.

The plutonium fuel is encased in various materials chosen for radiation shielding and for heat and shock resistance. The materials include a tantalum-tungsten alloy, a platinum-rhodium alloy, titanium, fibrous carbon, and graphite. The outside layer is stainless steel.

Extensive safety analyses and tests were performed by Sandia Laboratories at Albuquerque, New Mexico, to determine effects of an abort or any conceivable accident in connection with the Moon flight. The safety report by the Interagency Safety Evaluation Panel, which is made up of representatives of NASA, the AEC, and the Department of Defense, concluded that the heater presents no undue safety problem to the general population under any accident condition deemed possible for the Apollo mission.

-more-

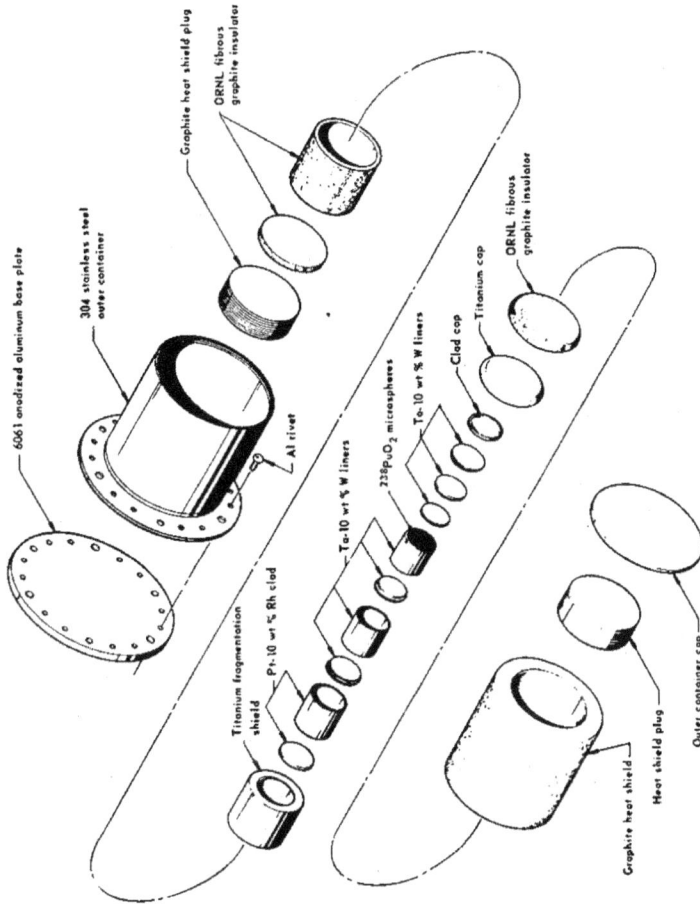

EXPLODED VIEW, APOLLO LUNAR RADIOISOTOPIC HEATER

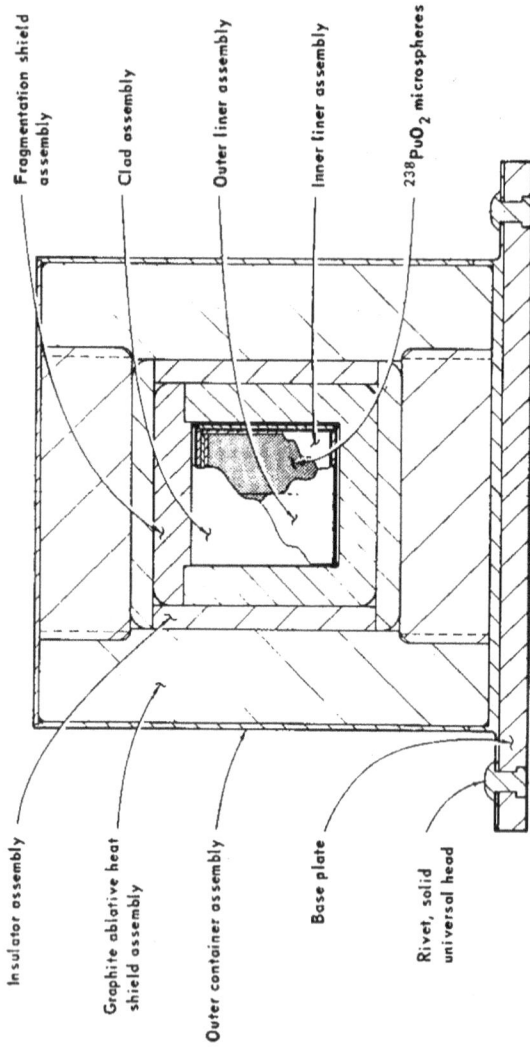

THE APOLLO LUNAR RADIOISOTOPIC HEATER

Fragmentation shield assembly

Clad assembly

Outer liner assembly

Inner liner assembly

$^{238}PuO_2$ microspheres

Insulator assembly

Graphite ablative heat shield assembly

Outer container assembly

Base plate

Rivet, solid universal head

Sandia Laboratories is operated for the AEC by Western Electric Company. The heater was fabricated by AEC's Mound Laboratory at Miamisburg, Ohio, which is operated by Monsanto Research Corporation.

The first major use of nuclear energy in space came in 1961 with the launching of a navigation satellite with an isotopic generator. Plutonium 238 fuels the device which is still operating. Two similar units were launched in 1961 and two more in 1963.

Last April, NASA launched Nimbus III, a weather satellite with a 2-unit nuclear isotopic system for generating electrical power. The Systems for Nuclear Auxiliary Power (SNAP-19) generator, developed by AEC, provides supplementary power.

Apollo 12 is scheduled to carry a SNAP-27 radioisotope thermoelectric generator, also developed by AEC, to provide power to operate the Apollo Lunar Surface Experiments Package (ALSEP). The SNAP-27 also contains plutonium 238 as the heat source. Thermoelectric elements convert this heat directly into electrical energy.

APOLLO LAUNCH OPERATIONS

Prelaunch Preparations

NASA's John F. Kennedy Space Center performs preflight checkout, test and launch of the Apollo 11 space vehicle. A government-industry team of about 500 will conduct the final countdown from Firing Room 1 of the Launch Control Center (LCC).

The firing room team is backed up by more than 5,000 persons who are directly involved in launch operations at KSC from the time the vehicle and spacecraft stages arrive at the Center until the launch is completed.

Initial checkout of the Apollo spacecraft is conducted in work stands and in the altitude chambers in the Manned Spacecraft Operations Building (MSOB) at Kennedy Space Center. After completion of checkout there, the assembled spacecraft is taken to the Vehicle Assembly Building (VAB) and mated with the launch vehicle. There the first integrated spacecraft and launch vehicle tests are conducted. The assembled space vehicle is then rolled out to the launch pad for final preparations and countdown to launch.

In early January, 1969, flight hardware for Apollo 11 began arriving at Kennedy Space Center, just as Apollo 9 and Apollo 10 were undergoing checkout at KSC.

The lunar module was the first piece of Apollo 11 flight hardware to arrive at KSC. The two stages of the LM were moved into the altitude chamber in the Manned Spacecraft Operations Building after an initial receiving inspection in January. In the chamber the LM underwent systems tests and both unmanned and manned chamber runs. During these runs the chamber air was pumped out to simulate the vacuum of space at altitudes in excess of 200,000 feet. There the spacecraft systems and the astronauts' life support systems were tested.

While the LM was undergoing preparation for its manned altitude chamber runs, the Apollo 11 command/service module arrived at KSC and after receiving inspection, it, too, was placed in an altitude chamber in the MSOB for systems tests and unmanned and manned chamber runs. The prime and backup crews participated in the chamber runs on both the LM and the CSM.

In early April, the LM and CSM were removed from the chambers. After installing the landing gear on the LM and the SPS engine nozzle on the CSM, the LM was encapsulated in the spacecraft LM adapter (SLA) and the CSM was mated to the SLA. On April 14, the assembled spacecraft was moved to the VAB where it was mated to the launch vehicle.

-more-

The launch vehicle flight hardware began arriving at KSC in mid-January and by March 5 the three stages and the instrument unit were erected on Mobile Launcher 1 in high bay 1. Tests were conducted on individual systems on each of the stages and on the overall launch vehicle before the spacecraft was erected atop the vehicle.

After spacecraft erection, the spacecraft and launch vehicle were electrically mated and the first overall test (plugs-in) of the space vehicle was conducted. In accordance with the philosophy of accomplishing as much of the checkout as possible in the VAB, the overall test was conducted before the space vehicle was moved to the launch pad.

The plugs-in test verified the compatibility of the space vehicle systems, ground support equipment and off-site support facilities by demonstrating the ability of the systems to proceed through a simulated countdown, launch and flight. During the simulated flight portion of the test, the systems were required to respond to both emergency and normal flight conditions.

The move to Pad A from the VAB on May 21 occurred while Apollo 10 was enroute to the Moon for a dress rehearsal of a lunar landing mission and the first test of a complete spacecraft in the near-lunar environment.

Apollo 11 will mark the fifth launch at Pad A on Complex 39. The first two unmanned Saturn V launches and the manned Apollo 8 and 9 launches took place at Pad A. Apollo 10 was the only launch to date from Pad B.

The space vehicle Flight Readiness Test was conducted June 4-6. Both the prime and backup crews participate in portions of the FRT, which is a final overall test of the space vehicle systems and ground support equipment when all systems are as near as possible to a launch configuration.

After hypergolic fuels were loaded aboard the space vehicle and the launch vehicle first stage fuel (RP-1) was brought aboard, the final major test of the space vehicle began. This was the countdown demonstration test (CDDT), a dress rehearsal for the final countdown to launch. The CDDT for Apollo 11 was divided into a "wet" and a "dry" portion. During the first, or "wet" portion, the entire countdown, including propellant loading, was carried out down to T-8.9 seconds, the time for ignition sequence start. The astronaut crew did not participate in the wet CDDT.

At the completion of the wet CDDT, the cryogenic propellants (liquid oxygen and liquid hydrogen) were off-loaded,
and the final portion of the countdown was re-run, this time
simulating the fueling and with the prime astronaut crew
participating as they will on launch day.

By the time Apollo 11 was entering the final phase of its
checkout procedure at Complex 39A, crews had already started
the checkout of Apollo 12 and Apollo 13. The Apollo 12 spacecraft completed altitude chamber testing in June and was later
mated to the launch vehicle in the VAB. Apollo 13 flight hardware began arriving in June to undergo preliminary checkout.

Because of the complexity involved in the checkout of the
363-foot-tall (110.6 meters) Apollo/Saturn V configuration, the
launch teams make use of extensive automation in their checkout.
Automation is one of the major differences in checkout used in
Apollo compared to the procedures used in the Mercury and
Gemini programs.

Computers, data display equipment and digital data techniques are used throughout the automatic checkout from the time
the launch vehicle is erected in the VAB through liftoff. A
similar, but separate computer operation called ACE (Acceptance
Checkout-Equipment) is used to verify the flight readiness of
the spacecraft. Spacecraft checkout is controlled from separate
rooms in the Manned Spacecraft Operations Building.

LAUNCH COMPLEX 39

Launch Complex 39 facilities at the Kennedy Space Center were planned and built specifically for the Apollo Saturn V, the space vehicle that will be used to carry astronauts to the Moon.

Complex 39 introduced the mobile concept of launch operations, a departure from the fixed launch pad techniques used previously at Cape Kennedy and other launch sites. Since the early 1950's when the first ballistic missiles were launched, the fixed launch concept had been used on NASA missions. This method called for assembly, checkout and launch of a rocket at one site--the launch pad. In addition to tying up the pad, this method also often left the flight equipment exposed to the outside influences of the weather for extended periods.

Using the mobile concept, the space vehicle is thoroughly checked in an enclosed building before it is moved to the launch pad for final preparations. This affords greater protection, a more systematic checkout process using computer techniques and a high launch rate for the future, since the pad time is minimal.

Saturn V stages are shipped to the Kennedy Space Center by ocean-going vessels and specially designed aircraft, such as the Guppy. Apollo spacecraft modules are transported by air. The spacecraft components are first taken to the Manned Spacecraft Operations Building for preliminary checkout. The Saturn V stages are brought immediately to the Vehicle Assembly Building after arrival at the nearby turning basin.

The major components of Complex 39 include: (1) the Vehicle Assembly Building (VAB) where the Apollo 11 was assembled and prepared; (2) the Launch Control Center, where the launch team conducts the preliminary checkout and final countdown; (3) the mobile launcher, upon which the Apollo 11 was erected for checkout and from where it will be launched; (4) the mobile service structure, which provides external access to the space vehicle at the pad; (5) the transporter, which carries the space vehicle and mobile launcher, as well as the mobile service structure to the pad; (6) the crawlerway over which the space vehicle travels from the VAB to the launch pad; and (7) the launch pad itself.

-more-

Vehicle Assembly Building

The Vehicle Assembly Building is the heart of Launch Complex 39. Covering eight acres, it is where the 363-foot-tall space vehicle is assembled and tested.

The VAB contains 129,482,000 cubic feet of space. It is 716 feet long, and 518 feet wide and it covers 343,500 square feet of floor space.

The foundation of the VAB rests on 4,225 steel pilings, each 16 inches in diameter, driven from 150 to 170 feet to bedrock. If placed end to end, these pilings would extend a distance of 123 miles. The skeletal structure of the building contains approximately 60,000 tons of structural steel. The exterior is covered by more than a million square feet of insulated aluminum siding.

The building is divided into a high bay area 525 feet high and a low bay area 210 feet high, with both areas serviced by a transfer aisle for movement of vehicle stages.

The low bay work area, approximately 442 feet wide and 274 feet long, contains eight stage-preparation and checkout cells. These cells are equipped with systems to simulate stage interface and operation with other stages and the instrument unit of the Saturn V launch vehicle.

After the Apollo 11 launch vehicle upper stages arrived at Kennedy Space Center, they were moved to the low bay of the VAB. Here, the second and third stages underwent acceptance and checkout testing prior to mating with the S-IC first stage atop the Mobile Launcher in the high bay area.

The high bay provides facilities for assembly and checkout of both the launch vehicle and spacecraft. It contains four separate bays for vertical assembly and checkout. At present, three bays are equipped, and the fourth will be reserved for possible changes in vehicle configuration.

Work platforms -- some as high as three-story buildings -- in the high bays provide access by surrounding the vehicle at varying levels. Each high bay has five platforms. Each platform consists of two bi-parting sections that move in from opposite sides and mate, providing a 360-degree access to the section of the space vehicle being checked.

A 10,000-ton-capacity air conditioning system, sufficient to cool about 3,000 homes, helps to control the environment within the entire office, laboratory, and workshop complex located inside the low bay area of the VAB. Air conditioning is also fed to individual platform levels located around the vehicle.

-more-

There are 141 lifting devices in the VAB, ranging from one-ton hoists to two 250-ton high-lift bridge cranes.

The mobile launchers, carried by transporter vehicles, move in and out of the VAB through four doors in the high bay area, one in each of the bays. Each door is shaped like an inverted T. They are 152 feet wide and 114 feet high at the base, narrowing to 76 feet in width. Total door height is 456 feet.

The lower section of each door is of the aircraft hangar type that slides horizontally on tracks. Above this are seven telescoping vertical lift panels stacked one above the other, each 50 feet high and driven by an individual motor. Each panel slides over the next to create an opening large enough to permit passage of the mobile launcher.

Launch Control Center

Adjacent to the VAB is the Launch Control Center (LCC). This four-story structure is a radical departure from the dome-shaped blockhouses at other launch sites.

The electronic "brain" of Launch Complex 39, the LCC was used for checkout and test operations while Apollo 11 was being assembled inside the VAB. The LCC contains display, monitoring, and control equipment used for both checkout and launch operations.

The building has telemeter checkout stations on its second floor, and four firing rooms, one for each high bay of the VAB, on its third floor. Three firing rooms contain identical sets of control and monitoring equipment, so that launch of a vehicle and checkout of others take place simultaneously. A ground computer facility is associated with each firing room.

The high speed computer data link is provided between the LCC and the mobile launcher for checkout of the launch vehicle. This link can be connected to the mobile launcher at either the VAB or at the pad.

The three equipped firing rooms have some 450 consoles which contain controls and displays required for the checkout process. The digital data links connecting with the high bay areas of the VAB and the launch pads carry vast amounts of data required during checkout and launch.

There are 15 display systems in each LCC firing room, with each system capable of providing digital information instantaneously.

Sixty television cameras are positioned around the Apollo/
Saturn V transmitting pictures on 10 modulated channels. The LCC
firing room also contains 112 operational intercommunication
channels used by the crews in the checkout and launch countdown.

Mobile Launcher

The mobile launcher is a transportable launch base and
umbilical tower for the space vehicle. Three mobile launchers are
used at Complex 39.

The launcher base is a two-story steel structure, 25 feet high,
160 feet long, and 135 feet wide. It is positioned on six steel
pedestals 22 feet high when in the VAB or at the launch pad. At
the launch pad, in addition to the six steel pedestals, four exten-
dable columns also are used to stiffen the mobile launcher against
rebound loads, if the Saturn engines cut off.

The umbilical tower, extending 398 feet above the launch platform, is mounted on one end of the launcher base. A hammerhead
crane at the top has a hook height of 376 feet above the deck with
a traverse radius of 85 feet from the center of the tower.

The 12-million-pound mobile launcher stands 445 feet high
when resting on its pedestals. The base, covering about half an
acre, is a compartmented structure built of 25-foot steel girders.

The launch vehicle sits over a 45-foot-square opening which
allows an outlet for engine exhausts into the launch pad trench
containing a flame deflector. This opening is lined with a re-
placeable steel blast shield, independent of the structure, and
is cooled by a water curtain initiated two seconds after liftoff.

There are nine hydraulically-operated service arms on the
umbilical tower. These service arms support lines for the vehicle
umbilical systems and provide access for personnel to the stages
as well as the astronaut crew to the spacecraft.

On Apollo 11, one of the service arms is retracted early in
the count. The Apollo spacecraft access arm is partially retracted at T-43 minutes. A third service arm is released at T-30
seconds, and a fourth at about T-16.5 seconds. The remaining
five arms are set to swing back at vehicle first motion after T-0.

The service arms are equipped with a backup retraction system
in case the primary mode fails.

The Apollo access arm (service arm 9), located at the 320-foot level above the launcher base, provides access to the spacecraft cabin for the closeout team and astronaut crews. The flight crew will board the spacecraft starting about T-2 hours, 40 minutes in the count. The access arm will be moved to a parked position, 12 degrees from the spacecraft, at about T-43 minutes. This is a distance of about three feet, which permits a rapid reconnection of the arm to the spacecraft in the event of an emergency condition. The arm is fully retracted at the T-5 minute mark in the count.

The Apollo 11 vehicle is secured to the mobile launcher by four combination support and hold-down arms mounted on the launcher deck. The hold-down arms are cast in one piece, about 6 x 9 feet at the base and 10 feet tall, weighing more than 20 tons. Damper struts secure the vehicle near its top.

After the engines ignite, the arms hold Apollo 11 for about six seconds until the engines build up to 95 percent thrust and other monitored systems indicate they are functioning properly. The arms release on receipt of a launch commit signal at the zero mark in the count. But the vehicle is prevented from accelerating too rapidly by controlled release mechanisms.

The mobile launcher provides emergency egress for the crew and closeout service personnel. Personnel may descend the tower via two 600-feet per minute elevators or by a slide-wire and cab to a bunker 2,200 feet from the launcher. If high speed elevators are utilized to level A of the launcher, two options are then available. The personnel may slide down the escape tube to the blast room below the pad or take elevator B to the bottom of the pad and board armored personnel carriers and depart the area.

Transporter

The six-million-pound transporters move mobile launchers into the VAB and mobile launchers with assembled Apollo space vehicles to the launch pad. They also are used to transfer the mobile service structure to and from the launch pads. Two transporters are in use at Complex 39.

The transporter is 131 feet long and 114 feet wide. The vehicle moves on four double-tracked crawlers, each 10 feet high and 40 feet long. Each shoe on the crawler track is seven feet six inches in length and weighs about a ton.

Sixteen traction motors powered by four 1,000-kilowatt generators, which in turn are driven by two 2,750-horsepower diesel engines, provide the motive power for the transporter. Two 750-kw generators, driven by two 1,065-horsepower diesel engines, power the jacking, steering, lighting, ventilating and electronic systems.

Maximum speed of the transporter is about one-mile-per-hour loaded and about two-miles-per-hour unloaded. The 3.5 mile trip to Pad A with Apollo 11 on its mobile launcher took about six hours since maximum speed is not maintained throughout the trip.

The transporter has a leveling system designed to keep the top of the space vehicle vertical within plus-or-minus 10 minutes of arc -- about the dimensions of a basketball.

This system also provides leveling operations required to negotiate the five percent ramp which leads to the launch pad and keeps the load level when it is raised and lowered on pedestals both at the pad and within the VAB.

The overall height of the transporter is 20 feet from ground level to the top deck on which the mobile launcher is mated for transportation. The deck is flat and about the size of a base-ball diamond (90 by 90 feet).

Two operator control cabs, one at each end of the chassis located diagonally opposite each other, provide totally enclosed stations from which all operating and control functions are coordinated.

Crawlerway

The transporter moves on a roadway 131 feet wide, divided by a median strip. This is almost as broad as an eight-lane turnpike and is designed to accommodate a combined weight of about 18 million pounds.

The roadway is built in three layers with an average depth of seven feet. The roadway base layer is two-and-one-half feet of hydraulic fill compacted to 95 percent density. The next layer consists of three feet of crushed rock packed to maximum density, followed by a layer of one foot of selected hydraulic fill. The bed is topped and sealed with an asphalt prime coat.

On top of the three layers is a cover of river rock, eight inches deep on the curves and six inches deep on the straightway. This layer reduces the friction during steering and helps distribute the load on the transporter bearings.

Mobile Service Structure

A 402-foot-tall, 9.8-million-pound tower is used to service the Apollo launch vehicle and spacecraft at the pad. The 40-story steel-trussed tower, called a mobile service structure, provides 360-degree platform access to the Saturn launch vehicle and the Apollo spacecraft.

The service structure has five platforms -- two self-propelled and three fixed, but movable. Two elevators carry personnel and equipment between work platforms. The platforms can open and close around the 363-foot space vehicle.

After depositing the mobile launcher with its space
vehicle on the pad, the transporter returns to a parking
area about 7,000 feet from pad A. There it picks up the
mobile service structure and moves it to the launch pad.
At the pad, the huge tower is lowered and secured to four
mount mechanisms.

The top three work platforms are located in fixed
positions which serve the Apollo spacecraft. The two lower
movable platforms serve the Saturn V.

The mobile service structure remains in position until
about T-11 hours when it is removed from its mounts and returned to the parking area.

Water Deluge System

A water deluge system will provide a million gallons
of industrial water for cooling and fire prevention during
launch of Apollo 11. Once the service arms are retracted at
liftoff, a spray system will come on to cool these arms from
the heat of the five Saturn F-1 engines during liftoff.

On the deck of the mobile launcher are 29 water nozzles.
This deck deluge will start immediately after liftoff and will
pour across the face of the launcher for 30 seconds at the rate
of 50,000 gallons-per-minute. After 30 seconds, the flow will
be reduced to 20,000 gallons-per-minute.

Positioned on both sides of the flame trench are a
series of nozzles which will begin pouring water at 8,000
gallons-per-minute, 10 seconds before liftoff. This water
will be directed over the flame deflector.

Other flush mounted nozzles, positioned around the pad,
will wash away any fluid spill as a protection against fire
hazards.

Water spray systems also are available along the
egress route that the astronauts and closeout crews would
follow in case an emergency evacuation was required.

Flame Trench and Deflector

The flame trench is 58 feet wide and approximately six
feet above mean sea level at the base. The height of the
trench and deflector is approximately 42 feet.

-more-

The flame deflector weighs about 1.3 million pounds and is stored outside the flame trench on rails. When it is moved beneath the launcher, it is raised hydraulically into position. The deflector is covered with a four-and-one-half-inch thickness of refractory concrete consisting of a volcanic ash aggregate and a calcium aluminate binder. The heat and blast of the engines are expected to wear about three-quarters of an inch from this refractory surface during the Apollo 11 launch.

Pad Areas

Both Pad A and Pad B of Launch Complex 39 are roughly octagonal in shape and cover about one fourth of a square mile of terrain.

The center of the pad is a hardstand constructed of heavily reinforced concrete. In addition to supporting the weight of the mobile launcher and the Apollo Saturn V vehicle, it also must support the 9.8-million-pound mobile service structure and 6-million-pound transporter, all at the same time. The top of the pad stands some 48 feet above sea level.

Saturn V propellants -- liquid oxygen, liquid hydrogen and RP-1 -- are stored near the pad perimeter.

Stainless steel, vacuum-jacketed pipes carry the liquid oxygen (LOX) and liquid hydrogen from the storage tanks to the pad, up the mobile launcher, and finally into the launch vehicle propellant tanks.

LOX is supplied from a 900,000-gallon storage tank. A centrifugal pump with a discharge pressure of 320 pounds-per-square-inch pumps LOX to the vehicle at flow rates as high as 10,000-gallons-per-minute.

Liquid hydrogen, used in the second and third stages, is stored in an 850,000-gallon tank, and is sent through 1,500 feet of 10-inch, vacuum-jacketed invar pipe. A vaporizing heat exchanger pressurizes the storage tank to 60 psi for a 10,000 gallons-per-minute flow rate.

The RP-1 fuel, a high grade of kerosene is stored in three tanks--each with a capacity of 86,000 gallons. It is pumped at a rate of 2,000 gallons-per-minute at 175 psig.

The Complex 39 pneumatic system includes a converter-compressor facility, a pad high-pressure gas storage battery, a high-pressure storage battery in the VAB, low and high-pressure, cross-country supply lines, high-pressure hydrogen storage and conversion equipment, and pad distribution piping to pneumatic control panels. The various purging systems require 187,000 pounds of liquid nitrogen and 21,000 gallons of helium.

-more-

Mission Control Center

The Mission Control Center at the Manned Spacecraft
Center, Houston, is the focal point for Apollo flight control
activities. The center receives tracking and telemetry data
from the Manned Space Flight Network, processes this data
through the Mission Control Center Real-Time Computer Complex,
and displays this data to the flight controllers and engineers
in the Mission Operations Control Room and staff support rooms.

The Manned Space Flight Network tracking and data
acquisition stations link the flight controllers at the center
to the spacecraft.

For Apollo 10 all network stations will be remote sites,
that is, without flight control teams. All uplink commands and
voice communications will originate from Houston, and telemetry
data will be sent back to Houston at high speed rates (2,400
bits-per-second), on two separate data lines. They can be
either real time or playback information.

Signal flow for voice circuits between Houston and
the remote sites is via commercial carrier, usually satellite,
wherever possible using leased lines which are part of the NASA
Communications Network.

Commands are sent from Houston to NASA's Goddard Space
Flight Center, Greenbelt, Md., on lines which link computers
at the two points. The Goddard communication computers provide automatic switching facilities and speed buffering for the
command data. Data are transferred from Goddard to remote sites
on high speed (2,400 bits-per-second) lines. Command loads also
can be sent by teletype from Houston to the remote sites at 100
words-per-minute. Again, Goddard computers provide storage and
switching functions.

Telemetry data at the remote site are received by
the RF receivers, processed by the pulse code modulation
ground stations, and transferred to the 642B remote-site
telemetry computer for storage. Depending on the format
selected by the telemetry controller at Houston, the 642B
will send the desired format through a 2010 data trans
mission unit which provides parallel to serial conversion,
and drives a 2,400 bit-per-second mode.

The data mode converts the digital serial data to
phase-shifted keyed tones which are fed to the high speed
data lines of the communications network.

Tracking data are sent from the sites in a low
speed (100 words) teletype format and a 240-bit block high
speed (2,400 bits) format. Data rates are one sample-6
seconds for teletype and 10 samples (frames) per second for
high speed data.

All high-speed data, whether tracking or telemetry,
which originate at a remote site are sent to Goddard on high-
speed lines. Goddard reformats the data when necessary and
sends them to Houston in 600-bit blocks at a 40,800 bits-per-
second rate. Of the 600-bit block, 480 bits are reserved for
data, the other 120 bits for address, sync, intercomputer instru-
ctions, and polynominal error encoding.

All wideband 40,800 bits-per-second data originating at
Houston are converted to high speed (2,400 bits-per-second)
data at Goddard before being transferred to the designated
remote site.

-more-

MANNED SPACE FLIGHT NETWORK

Tracking, command and communication -- Apollo 11's vital links with the Earth -- will be performed, in two broad phases.

For the first phase, the Manned Space Flight Network (MSFN) will depend largely on its worldwide chain of stations equipped with 30-foot antennas while Apollo is launched and orbiting near the Earth. The second phase begins when the spacecraft moves out more than 10,000 miles above Earth, when the 85-foot diameter antennas bring their greater power and accuracy into play.

The Network must furnish reliable, instantaneous contact with the astronauts, their launch vehicle and spacecraft, from liftoff through Earth orbit, Moon landing and lunar takeoff to splashdown in the Pacific Ocean.

For Apollo 11, MSFN will use 17 ground stations, four ships and six to eight jet aircraft -- all directly or indirectly linked with Mission Control Center in Houston. While the Earth turns on its axis and the Moon travels in orbit nearly one-quarter million miles away and Apollo 11 moves between them, ground controllers will be kept in the closest possible contact. Thus, only for some 45 minutes as the spacecraft flies behind the Moon in each orbit, will this link with Earth be out of reach.

All elements of the Network get ready early in the countdown. As the Apollo Saturn V ascends, voice and data will be transmitted instantaneously to Houston. The data are sent directly through computers for visual display to flight controllers.

Depending on the launch azimuth, the 30-foot antennas will keep tabs on Apollo 11, beginning with the station at Merritt Island, thence Grand Bahama Island; Bermuda; tracking ship Vanguard; the Canary Islands; Carnarvon, Australia; Hawaii; another tracking ship; Guaymas, Mexico; and Corpus Christi, Tex.

To inject Apollo 11 into translunar flight path, Mission Control will send a signal through one of the land stations or one of the tracking ships in the Pacific. As the spacecraft heads for the Moon, the engine burn will be monitored by the ships and an Apollo range instrumentation aircraft (ARIA). The ARIA provides a relay for the astronauts' voices and data communication with Houston.

MANNED SPACE FLIGHT TRACKING NETWORK

When the spacecraft reaches an altitude of 10,000 miles the more powerful 85-foot antennas will join in for primary support of the flight, although the 30-foot "dishes" will continue to track and record data. The 85-foot antennas are located, about 120 degrees apart, near Madrid, Spain; Goldstone, Calif.; and Canberra, Australia.

With the 120-degree spacing around the Earth, at least one of the large antennas will have the Moon in view at all times. As the Earth revolves from west to east, one 85-foot station hands over control to the next 85-foot station as it moves into view of the spacecraft. In this way, data and communication flow is maintained.

Data are relayed back through the huge antennas and transmitted via the NASA Communications Network (NASCOM) -- a two-million mile hookup of landlines, undersea cables, radio circuits and communication satellites -- to Houston. This informatin is fed into computers for visual display in Mission Control -- for example, a display of the precise position of the spacecraft on a large map. Or, returning data my indicate a drop in power or some other difficulty in a spacecraft system, which would energize a red light to alert a flight controller to action.

Returning data flowing through the Earth stations give the necessary information for commanding midcourse maneuvers to keep Apollo 11 in a proper trajectory for orbiting the Moon. After Apollo 11 is in the vicinity of the Moon, these data indicate the amount of retro burn necessary for the service module engine to place the spacecraft in lunar orbit.

Once the lunar module separates from the command module and goes into a separate lunar orbit, the MSFN will be required to keep track of both spacecraft at once, and provide two-way communication and telemetry between them and the Earth. The prime antenna at each of the three 85-foot tracking stations will handle one spacecraft while a wing, or backup, antenna at the same site will handle the other spacecraft during each pass.

Tracking and acquisition of data between Earth and the two spacecraft will provide support for the rendezvous and docking maneuvers. The information will also be used to determine the time and duration of the service module propulsion engine burn required to place the command module into a precise trajectory for reentering the Earth's atmosphere at the planned location.

As the spacecraft comes toward Earth at high speed -- up to more than 25,000 miles per hour -- it must reenter at the proper angle. To make an accurate reentry, information from the tracking stations and ships is fed into the MCC computers where flight controllers make decisions that will provide the Apollo 11 crew with the necessary information.

Appropriate MSFN stations, including the ships and aircraft in the Pacific, are on hand to provide support during the reentry. An ARIA aircraft will relay astronaut voice communications to MCC and antennas on reentry ships will follow the spacecraft.

Through the journey to the Moon and return, television will be received from the spacecraft at the three 85-foot antennas around the world. In addition, 210-foot diameter antennas in California and Australia will be used to augment the television coverage while the Apollo 11 is near and on the Moon. Scan converters at the stations permit immediate transmission of commercial quality TV via NASCOM to Houston, where it will be released to TV networks.

NASA Communications Network

The NASA Communications Network (NASCOM) consists of several systems of diversely routed communications channels leased on communications satellites, common carrier systems and high frequency radio facilities where necessary to provide the access links.

The system consists of both narrow and wide-band channels, and some TV channels. Included are a variety of telegraph, voice, and data systems (digital and analog) with several digital data rates. Wide-band systems do not extend overseas. Alternate routes or redundancy provide added reliability.

A primary switching center and intermediate switching and control points provide centralized facility and technical control, and switching operations under direct NASA control. The primary switching center is at the Goddard Space Flight Center, Greenbelt, Md. Intermediate switching centers are located at Canberra, Madrid, London, Honolulu, Guam, and Kennedy Space Center.

For Apollo 11, the Kennedy Space Center is connected directly to the Mission Control Center, Houston via the Apollo Launch Data System and to the Marshall Space Flight Center, Huntsville, Ala., by a Launch Information Exchange Facility.

After launch, all network tracking and telemetry data hubs at GSFC for transmission to MCC Houston via two 50,000 bits-per-second circuits used for redundancy and in case of data overflow.

NASA COMMUNICATIONS NETWORK

Two Intelsat communications satellites will be used for Apollo 11. The Atlantic satellite will service the Ascension Island unified S-band (USB) station, the Atlantic Ocean ship and the Canary Islands site.

The second Apollo Intelsat communications satellite over the mid-Pacific will service the Carnarvon, Australia USB site and the Pacific Ocean ships. All these stations will be able to transmit simultaneously through the satellite to Houston via Brewster Flat, Wash., and the Goddard Space Flight Center, Greenbelt. Md.

Network Computers

At fraction-of-a-second intervals, the network's digital data processing systems, with NASA's Manned Spacecraft Center as the focal point, "talk" to each other or to the spacecraft. High-speed computers at the remote site (tracking ships included) issue commands or "up-link" data on such matters as control of cabin pressure, orbital guidance commands, or "go-no-go" indications to perform certain functions.

When information originates from Houston, the computers refer to their pre-programmed information for validity before transmitting the required data to the spacecraft.

Such "up-link"information is communicated by ultra-high-frequency radio about 1,200 bits-per-second. Communication between remote ground sites, via high-speed communications links, occurs at about the same rate. Houston reads information from these ground sites at 2,400 bits-per-second, as well as from remote sites at 100 words-per-minute.

The computer systems perform many other functions, including:

. Assuring the quality of the transmission lines by continually exercising data paths.

. Verifying accuracy of the messages by repetitive operations.

. Constantly updating the flight status.

For "down link" data, sensors built into the spacecraft continually sample cabin temperature, pressure, physical information on the astronauts such as heartbeat and respiration, among other items. These data are transmitted to the ground stations at 51.2 kilobits (12,800 binary digits) per-second.

At MCC the computers:

.   Detect and select changes or deviations, compare with
    their stored programs, and indicate the problem areas
    or pertinent data to the flight controllers.

.   Provide displays to mission personnel.

.   Assemble output data in proper formats.

.   Log data on magnetic tape for replay for the flight
    controllers.

.   Keep time.

The Apollo Ships

The mission will be supported by four Apollo instrumentation ships operating as integral stations of the Manned Space Flight Network (MSFN) to provide coverage in areas beyond the range of land stations.

The ships, USNS Vanguard, Redstone, Mercury, and Huntsville will perform tracking, telemetry, and communication functions for the launch phase, Earth orbit insertion, translunar injection, and reentry.

Vanguard will be stationed about 1,000 miles southeast of Bermuda (25 degrees N., 49 degrees W.) to bridge the Bermuda-Antigua gap during Earth orbit insertion. Vanguard also functions as part of the Atlantic recovery fleet in the event of a launch phase contingency. Redstone (2.25 degrees S., 166.8 degrees E.); the Mercury (10 N., 175.2 W.) and the Huntsville (3.0 N., 154.0 E.) provide a triangle of mobile stations between the MSFN stations at Carnarvon and Hawaii for coverage of the burn interval for translunar injection. In the event the launch date slips from July 16, the ships will all move generally northeastward to cover the changing translunar injection location.

Redstone and Huntsville will be repositioned along the reentry corridor for tracking, telemetry, and communications functions during reentry and landing. They will track Apollo from about 1,000 miles away through communications blackout when the spacecraft will drop below the horizon and will be picked up by the ARIA aircraft.

The Apollo ships were developed jointly by NASA and the Department of Defense. The DOD operates the ships in support of Apollo and other NASA and DOD missions on a non-interference basis with Apollo requirements.

Management of the Apollo ships is the responsibility of the Commander, Air Force Western Test Range (AFWTR). The Military Sea Transport Service provides the maritime crews and the Federal Electric Corp., International Telephone and Telegraph, under contract to AFWTR, provides the technical instrumentation crews.

The technical crews operate in accordance with joint NASA/DOD standards and specifications which are compatible with MSFN operational procedures.

Apollo Range Instrumentation Aircraft (ARIA)

During Apollo 11, the ARIA will be used primarily to fill coverage gaps between the land and ship stations in the Pacific between Australia and Hawaii during the translunar injection interval. Prior to and during the burn, the ARIA record telemetry data from Apollo and provide realtime voice communication between the astronauts and the Mission Control Center at Houston.

Eight aircraft will participate in this mission, operating from Pacific, Australian and Indian Ocean air fields in positions under the orbital track of the spacecraft and launch vehicle. The aircraft will be deployed in a northwestward direction in the event of launch day slips.

For reentry, two ARIA will be deployed to the landing area to continue communications between Apollo and Mission Control at Houston and provide position information on the spacecraft after the blackout phase of reentry has passed.

The total ARIA fleet for Apollo missions consists of eight EC-135A (Boeing 707) jets equipped specifically to meet mission needs. Seven-foot parabolic antennas have been installed in the nose section of the planes giving them a large, bulbous look.

The aircraft, as well as flight and instrumentation crews, are provided by the Air Force and they are equipped through joint Air Force-NASA contract action to operate in accordance with MSFN procedures.

Ship Positions for Apollo 11

July 16, 1969

Insertion Ship (VAN)	25 degrees N 49 degrees W
Injection Ship (MER)	10 degrees N 175.2 degrees W
Injection Ship (RED)	2.25 degrees S 166.8 degrees E
Injection Ship (HTV)	3.0 degrees N 154.0 degrees E
Reentry Support	
Reentry Ship (HTV)	5.5 degrees N 178.2 degrees W
Reentry Ship (RED)	3.0 degrees S 165.5 degrees E

July 18, 1969

Insertion Ship (VAN)	25 degrees N 49 degrees W
Injection Ship (MER)	15 degrees N 166.5 degrees W
Injection Ship (RED)	4.0 degrees N 172.0 degrees E
Injection Ship (HTV)	10.0 degrees N 157.0 degrees E
Reentry Support	
Reentry Ship (HTV)	17.0 degrees N 177.3 degrees W
Reentry Ship (RED)	6.5 degrees N 163.0 degrees E

July 21, 1969

Insertion Ship (VAN)	25 degrees N 49 degrees W
Injection Ship (MER)	16.5 degrees N 151 degrees W
Injection Ship (RED)	11.5 degrees N 177.5 degrees W
Injection Ship (HTV)	12.0 degrees N 166.0 degrees E
Reentry Support	
Reentry Ship (HTV)	26.0 degrees N 176.8 degrees W
Reentry Ship (RED)	17.3 degrees N 160.0 degrees E

-more-

CONTAMINATION CONTROL PROGRAM

In 1966 an Interagency Committee on Back Contamination (ICBC) was established. The function of this Committee was to assist NASA in developing a program to prevent the contamination of the Earth from lunar materials following manned lunar exploration. The committee charter included specific authority to review and approve the plans and procedures to prevent back contamination. The committee membership includes representatives from the Public Health Service, Department of Agriculture, Department of the Interior, NASA, and the National Academy of Sciences.

Over the last several years NASA has developed facilities, equipment and operational procedures to provide an adequate back contamination program for the Apollo missions. This program of facilities and procedures, which is well beyond the current state-of-the-art, and the overall effort have resulted in a laboratory with capabilities which have never previously existed. The scheme of isolation of the Apollo crewmen and lunar samples, and the exhaustive test programs to be conducted are extensive in scope and complexity.

The Apollo Back Contamination Program can be divided into three phases. The first phase covers the procedures which are followed by the crew while in flight to reduce and, if possible, eliminate the return of lunar surface contaminants in the command module.

The second phase includes spacecraft and crew recovery and the provisions for isolation and transport of the crew, spacecraft, and lunar samples to the MannedSpacecraft Center. The third phase encompasses the quarantine operations and preliminary sample analysis in the Lunar Receiving Laboratory.

A primary step in preventing back contamination is careful attention to spacecraft cleanliness following lunar surface operations. This includes use of special cleaning equipment, stowage provisions for lunar-exposed equipment, and crew procedures for proper "housekeeping."

Lunar Module Operations - The lunar module has been designed with a bacterial filter system to prevent contamination of the lunar surface when the cabin atmosphere is released at the start of the lunar exploration.

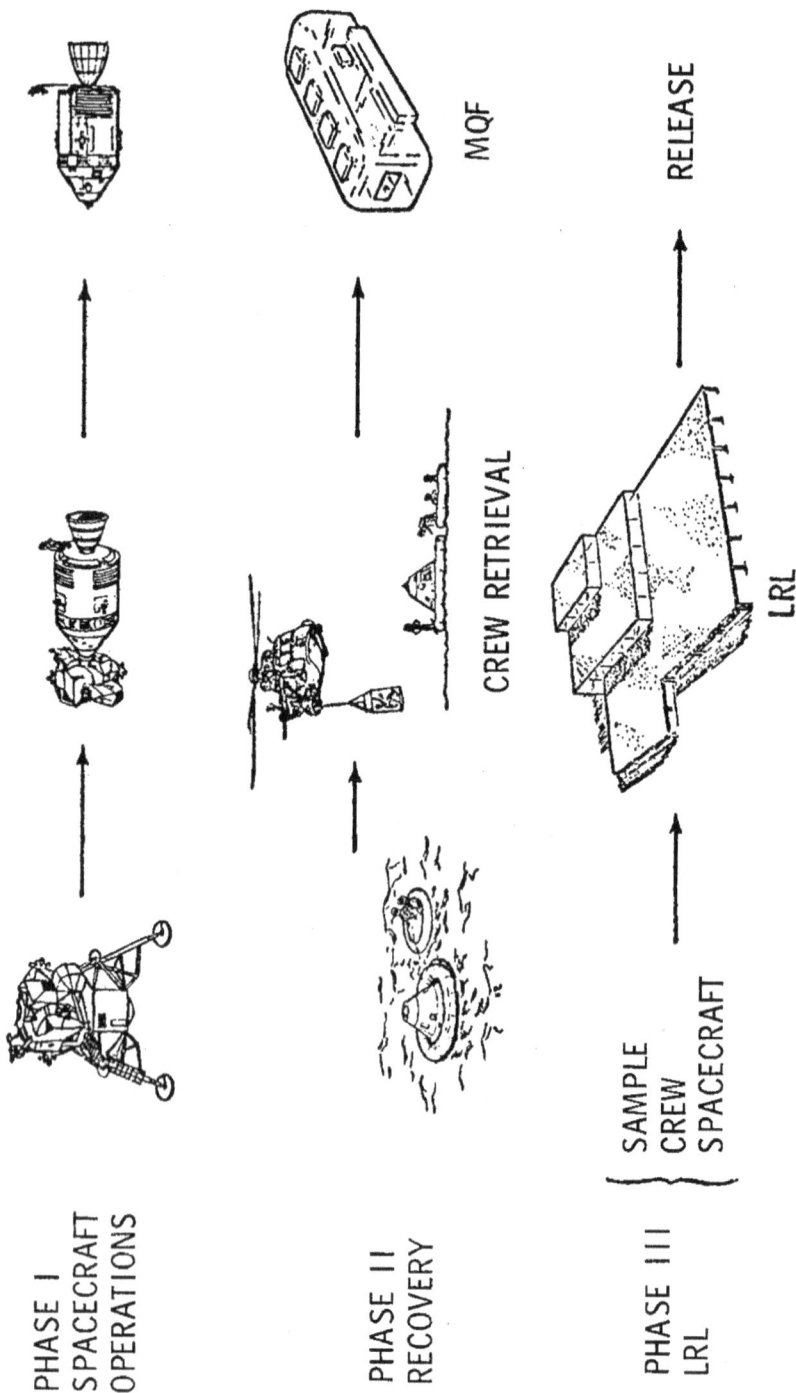

APOLLO BACK CONTAMINATION PROGRAM

PHASE I
SPACECRAFT
OPERATIONS

PHASE II
RECOVERY

CREW RETRIEVAL

MQF

PHASE III
LRL

SAMPLE
CREW
SPACECRAFT

LRL

RELEASE

Prior to reentering the LM after lunar surface exploration, the crewmen will brush any lunar surface dust or dirt from the space suit using the suit gloves. They will scrape their overboots on the LM footpad and while ascending the LM ladder dislodge any clinging particles by a kicking action.

After entering the LM and pressurizing the cabin, the crew will doff their portable life support system, oxygen purge system, lunar boots, EVA gloves, etc.

The equipment shown in Table I as jettisoned equipment will be assembled and bagged to be subsequently left on the lunar surface. The lunar boots, likely the most contaminated items, will be placed in a bag as early as possible to minimize the spread of lunar particles.

Following LM rendezvous and docking with the CM, the CM tunnel will be pressurized and checks made to insure that an adequate pressurized seal has been made. During this period, the LM, space suits, and lunar surface equipment will be vacuumed. To accomplish this, one additional lunar orbit has been added to the mission.

The lunar module cabin atmosphere will be circulated through the environmental control system suit circuit lithium hydroxide (LiOH) canister to filter particles from the atmosphere. A minimum of five hours weightless operation and filtering will reduce the original airborne contamination to about 10^{-15} per cent.

To prevent dust particles from being transferred from the LM atmosphere to the CM, a constant flow of 0.8 lbs/hr oxygen will be initiated in the CM at the start of combined LM/CM operation. Oxygen will flow from the CM into the LM then overboard through the LM cabin relief valve or through spacecraft leakage. Since the flow of gas is always from the CM to the LM, diffusion and flow of dust contamination into the CM will be minimized. After this positive gas flow has been established from the CM, the tunnel hatch will be removed.

The CM pilot will transfer the lunar surface equipment stowage bags into the LM one at a time. The equipment listed in Table 1 as equipment transferred will then be bagged using the "Buddy System" and transferred back into the CM where the equipment will be stowed. The only equipment which will not be bagged at this time are the crewmen's space suits and flight logs.

LUNAR SURFACE EQUIPMENT - CLEANING AND TRANSFER

Table I

ITEM	LOCATION AFTER JETTISON	EQUIPMENT LOCATION AT LUNAR LAUNCH	LM-CM TRANSFER	REMARKS
Jettisoned Equipment:				
Overshoes (In Container)	Lunar surface			
Portable Life Support System	"			
Camera	"			
Lunar tool tether	"			
Spacesuit connector cover	"			
Equipment Left in LM:				
EVA tether	RH side stowage container	RH side stowage container		Equipment Brushed prior to stowage for launch
EVA visors	Helmet bag	Helmet bag		
EVA gloves	Helmet bag	Helmet bag		
Purge valve	Interim stowage assy	Interim stowage assy		
Oxygen purge system	Engine cover	Engine cover		
Equipment Transferred to CM:				
Spacesuit	On crew	On crew	Stowed in bag	All equipment to be cleaned by vacuum brush prior to transfer to CM
Liquid-cooled garment	On crew	On crew	On crew	
Helmet	On crew	On crew	Stowed in bag	
Watch	On crew	On crew	On crew	
Lunar grab sample	LH stowage	LH stowage	Stowed in bag	
Lunar sample box	SRC rack	SRC rack	Stowed in bag	
Film magazine	SRC rack	SRC rack	Stowed in bag	

POSITIVE GAS FLOW
FROM CM TO LM AFTER POSTLANDING DOCKING

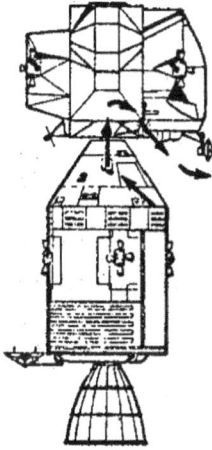

- **PROCEDURES**
  - PRESSURIZE TUNNEL
  - CM CABIN PRESSURE RELIEF VALVES POSITIONED TO CLOSED
  - LM FORWARD HATCH DUMP/RELIEF VALVE VERIFIED IN AUTOMATIC
  - CM DIRECT O_2 VALVE OPENED TO ESTABLISH CM CABIN PRESSURE AT LEAST 0.5 PSI GREATER THAN LM
  - OPEN PRESSURE EQUALIZATION ON TUNNEL HATCH
  - OBSERVE LM CABIN PRESSURE RELIEF FUNCTION
  - ADJUST CM DIRECT O_2 TO STABLE 0.8 #/HR
  - OPEN TUNNEL HATCH

OXYGEN USAGE RATES FOR POSITIVE GAS FLOW FROM CM TO LM

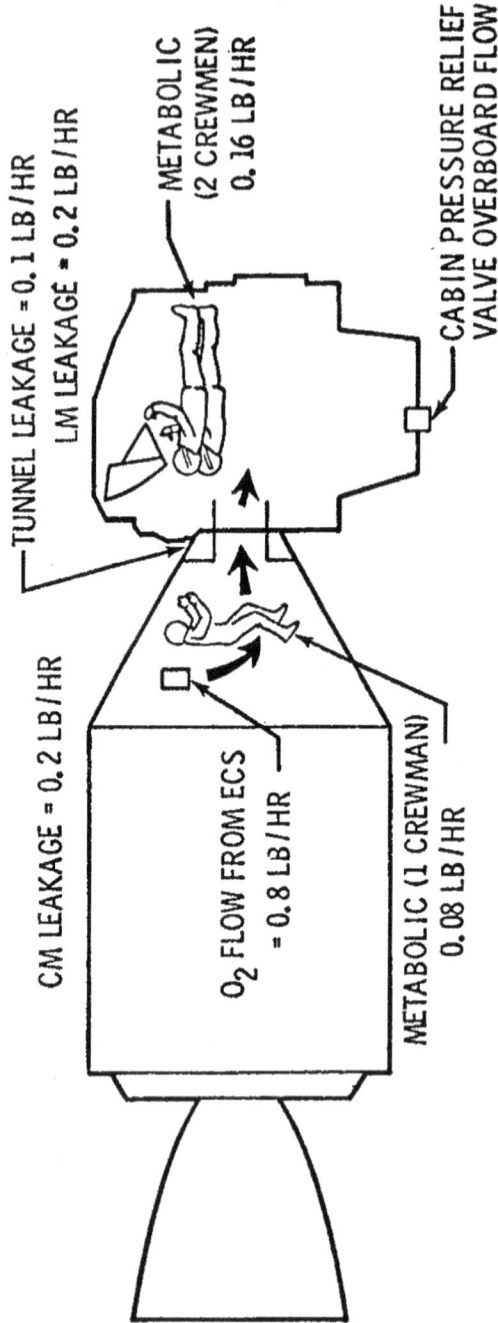

TUNNEL LEAKAGE = 0.1 LB/HR
LM LEAKAGE = 0.2 LB/HR

METABOLIC (2 CREWMEN) 0.16 LB/HR

CABIN PRESSURE RELIEF VALVE OVERBOARD FLOW

CM LEAKAGE = 0.2 LB/HR

O_2 FLOW FROM ECS = 0.8 LB/HR

METABOLIC (1 CREWMAN) 0.08 LB/HR

NOMINAL OXYGEN USAGE RATES

CM METABOLIC RATE	0.08 LB/HR
CM LEAKAGE	0.20 LB/HR
TUNNEL LEAKAGE	0.10 LB/HR
LM METABOLIC RATE	0.16 LB/HR
LM LEAKAGE	0.20 LB/HR
FLOW THRU LM CABIN PRESSURE RELIEF VALVE	0.06 LB/HR

Following the transfer of the LM crew and equipment, the spacecraft will be separated and the three crewmen will start the return to Earth. The separated LM contains the remainder of the lunar exposed equipment listed in Table 1.

Command Module Operations - through the use of operational and housekeeping procedures the command module cabin will be purged of lunar surface and/or other particulate contamination prior to Earth reeentry. These procedures start while the LM is docked with the CM and continue through reentry into the Earth's atmosphere.

The LM crewmen will doff their space suits immediately upon separation of the LM and CM. The space suits will be stowed and will not be used again during the trans-Earth phase unless an emergency occurs.

Specific periods for cleaning the spacecraft using the vacuum brush have been established. Visible liquids will be removed by the liquid dump system. Towels will be used by the crew to wipe surfaces clean of liquids and dirt particles. The three ECS suit hoses will be located at random positions around the spacecraft to insure positive ventilation, cabin atmosphere filtration, and avoid partitioning.

During the transearth phase, the command module atmosphere will be continually filtered through the environmental control system lithium hydroxide canister. This will remove essentially all airborne dust particles. After about 63 hours operation essentially none (10-90 per cent) of the original contaminates will remain.

Lunar Mission Recovery Operations

Following landing and the attachment of the flotation collar to the command module, the swimmer in a biological isolation garment (BIG) will open the spacecraft hatch, pass three BIGs into the spacecraft, and close the hatch.

The crew will don the BIGs and then egress into a life raft containing a decontaminant solution. The hatch will be closed immediately after egress. Tests have shown that the crew can don their BIGs in less than 5 minutes under ideal sea conditions. The spacecraft hatch will only be open for a matter of a few minutes. The spacecraft and crew will be decontaminated by the swimmer using a liquid agent.

Crew retrieval will be accomplished by helicopter to the carrier and subsequent crew transfer to the Mobile Quarantine Facility. The spacecraft will be retrieved by the aircraft carrier.

Biological Isolation Garment - Biological isolation garment (BIGs), will be donned in the CM just prior to egress and helicopter pick-up and will be worn until the crew enters the Mobile Quarantine Fac-ility aboard the primary recovery ship.

The suit is fabricated of a light weight cloth fabric which completely covers the wearer and serves as a biological barrier. Built into the hood area is a face mask with a plastic visor, air inlet flapper valve, and an air outlet biological filter.

Two types of BIGs are used in the recovery operation. One is worn by the recovery swimmer. In this type garment, the inflow air (inspired) is filtered by a biological filter to preclude possible contamination of support personnel. The second type is worn by the astronauts. The inflow gas is not filtered, but the outflow gas (respired) is passed through a biological flter to preclude contamination of the air.

Mobile Quarantine Facility - The Mobile Quarantine Facility, is equipped to house six people for a period up to 10 days. The interior is divided into three sections--lounge area, galley, and sleep/bath area. The facility is powered through several systems to interface with various ships, aircraft, and transportation vehicles. The shell is air and water tight. The principal method of assuring quarantine is to filter effluent air and provide a negative pressure differential for biological containment in the event of leaks.

Non-fecal liquids from the trailer are chemically treated and stored in special containers. Fecal wastes will be contained until after the quarantine period. Items are passed in or out of the MQF through a submersible transfer lock. A complete communications system is provided for intercom and external communications to land bases from ship or aircraft. Emergency alarms are provided for oxygen alerts while in transport by aircraft for fire, loss of power and loss of negative pressure.

Specially packaged and controlled meals will be passed into the facility where they will be prepared in a micro-wave oven. Medical equipment to complete immediate postlanding crew examination and tests are provided.

Lunar Receiving Laboratory - The final phase of the back contamination program is completed in the MSC Lunar Receiving Laboratory. The crew and spacecraft are quarantined for a minimum of 21 days after lunar liftoff and are released based upon the completion of prescribed test requirements and results. The lunar sample will be quarantined for a period of 50 to 80 days depending upon the result of extensive biological tests.

The LRL serves four basic purposes:

The quarantine of the lunar mission crew and spacecraft, the containment of lunar and lunar-exposed materials and quarantine testing to search for adverse effects of lunar material upon terrestrial life.

The preservation and protection of the lunar samples.

The performance of time critical investigations.

The preliminary examination of returned samples to assist in an intelligent distribution of samples to principal investigators.

The LRL has the only vacuum system in the world with space gloves operated by a man leading directly into a vacuum chamber at pressures of 10^{-7} torr. (mm Hg). It has a low level counting facility, whose background count is an order of magnitude better than other known counters. Additionally, it is a facility that can handle a large variety of biological specimens inside Class III biological cabinets designed to contain extremely hazardous pathogenic material.

The LRL, covers 83,000 square feet of floor space and includes several distinct areas. These are the Crew Reception Area (CRA), Vacuum Laboratory, Sample Laboratories (Physical and Bio-Science) and an administrative and support area. Special building systems are employed to maintain air flow into sample handling areas and the CRA to sterilize liquid waste and to incinerate contaminated air from the primary containment systems.

The biomedical laboratories provide for the required quarantine tests to determine the effect of lunar samples on terrestrial life. These tests are designed to provide data upon which to base the decision to release lunar material from quarantine.

Among the tests:

a. Germ-free mice will be exposed to lunar material and observed continuously for 21 days for any abnormal changes. Periodically, groups will be sacrificed for pathologic observation.

b. Lunar material will be applied to 12 different culture media and maintained under several environmental conditions. The media will then be observed for bacterial or fungal growth. Detailed inventories of the microbial flora of the spacecraft and crew have been maintained so that any living material found in the sample testing can be compared against this list of potential contaminants taken to the Moon by the crew or spacecraft.

c. Six types of human and animal tissue culture cell lines will be maintained in the laboratory and together with embryonated eggs are exposed to the lunar material. Based on cellular and/or other changes, the presence of viral material can be established so that special tests can be conducted to identify and isolate the type of virus present.

d. Thirty-three species of plants and seedlings will be exposed to lunar material. Seed germination, growth of plant cells or the health of seedlings then observed, and histological, microbiological and biochemical techniques used to determine the cause of any suspected abnormality.

e. A number of lower animals will be exposed to lunar material. These specimens include fish, birds, oysters, shrimp, cockroaches, houseflies, planaria, paramecia and euglena. If abnormalities are noted, further tests will be conducted to determine if the condition is transmissible from one group to another.

The crew reception area provides biological containment for the flight crew and 12 support personnel. The nominal occupancy is about 14 days but the facility is designed and equipped to operate for considerably longer if necessary.

Sterilization And Release Of The Spacecraft

Postflight testing and inspection of the spacecraft is presently limited to investigation of anomalies which happened during the flight. Generally, this entails some specific testing of the spacecraft and removal of certain components of systems for further analysis. The timing of postflight testing is important so that corrective action may be taken for subsequent flights.

The schedule calls for the spacecraft to be returned to port where a team will deactivate pyrotechnics, flush and drain fluid systems (except water). This operation will be confined to the exterior of the spacecraft. The spacecraft will then be flown to the LRL and placed in a special room for storage, sterilization, and postflight checkout.

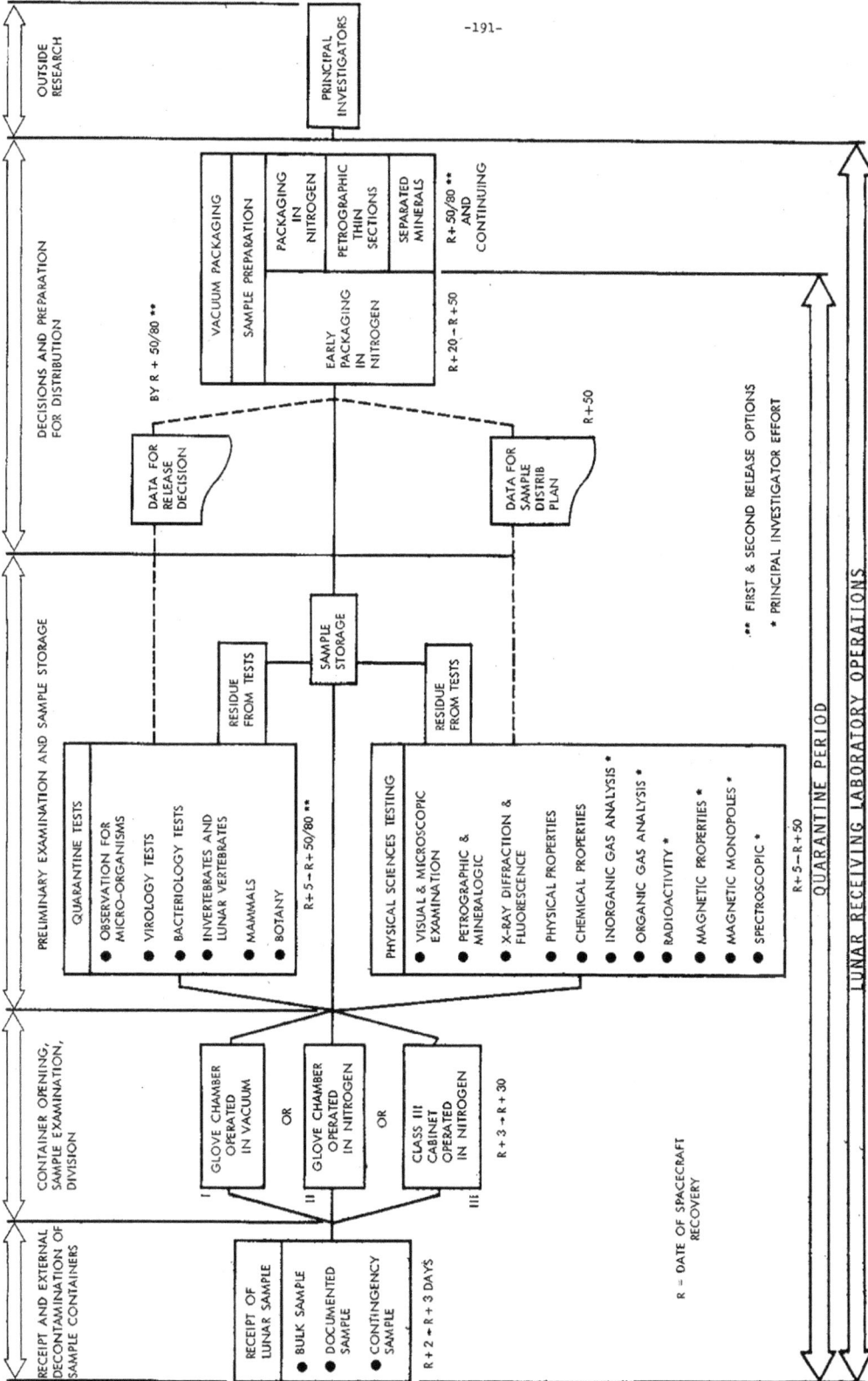

LUNAR SAMPLE OPERATIONS

APOLLO PROGRAM MANAGEMENT

The Apollo Program, the United States effort to land men on the Moon and return them safely to Earth before 1970, is the responsibility of the Office of Manned Space Flight (OMSF), National Aeronautics and Space Administration, Washington, D.C. Dr. George E. Mueller is Associate Administator for Manned Space Flight.

NASA Manned Spacecraft Center (MSC), Houston, is responsible for development of the Apollo spacecraft, flight crew training and flight control. Dr. Robert R. Gilruth is Center Director.

NASA Marshall Space Flight Center (MSFC), Huntsville, Ala., is responsible for development of the Saturn launch vehicles. Dr. Wernher von Braun is Center Director.

NASA John F. Kennedy Space Center (KSC), Fla., is responsible for Apollo/Saturn launch operations. Dr. Kurt H. Debus is Center Director.

The NASA Office of Tracking and Data Acquisition (OTDA) directs the program of tracking and data flow on Apollo 11. Gerald M. Truszynski is Associate Administrator for Tracking and Data Acquisition.

NASA Goddard Space Flight Center (GSFC), Greenbelt, Md., manages the Manned Space Flight Network (MSFN) and Communications Network (NASCOM). Dr. John F. Clark is Center Director.

The Department of Defense is supporting NASA in Apollo 11 during launch, tracking and recovery operations. The Air Force Eastern Test Range is responsible for range activities during launch and down-range tracking. DOD developed jointly with NASA the tracking ships and aircraft. Recovery operations include the use of recovery ships and Navy and Air Force aircraft.

Apollo/Saturn Officials

NASA Headquarters

DR. THOMAS O. PAINE was appointed NASA Administrator March 5, 1969. He was born in Berkeley, Calif., Nov. 9, 1921. Dr. Paine was graduated from Brown University in 1942 with an A.B. degree in engineering. After service as a submarine officer during World War II, he attended Stanford University, receiving an M.S. degree in 1947 and Ph.D. in 1949 in physical metallurgy. Dr. Paine worked as research associate at Stanford from 1947 to 1949 when he joined the General Electric Research Laboratory, Schenectady, N.Y. In 1951 he transferred to the Meter and Instrument Department, Lynn, Mass., as Manager of Materials Development, and later as laboratory manager. From 1958 to 1962 he was research associate and manager of engineering applications at GE's Research and Development Center in Schenectady. In 1963-68 he was manager of TEMPO, GE's Center for Advanced Studies in Santa Barbara, Calif.

On January 31, 1968, President Johnson appointed Dr. Paine Deputy Administrator of NASA, and he was named Acting Administrator upon the retirement of Mr. James E. Webb on Oct. 8, 1968. His nomination as Administrator was announced by President Nixon on March 5, 1969; this was confirmed by the Senate on March 20, 1969. He was sworn in by Vice President Agnew on April 3, 1969.

* * *

LIEUTENANT GENERAL SAMUEL C. PHILLIPS director of the United States Apollo Lunar Landing Program, was born in Arizona in 1921 and at an early age he moved to Cheyenne, Wyoming which he calls his permanent home. He graduated from the University of Wyoming in 1942 with a B.S. degree in electrical engineering and a presidential appointment as a second lieutenant of infantry in the regular army. He transferred to the Air Corps and earned his pilot's wings in 1943. Following wartime service as a combat pilot in Europe, he studied at the University of Michigan where he received his master of science degree in electrical engineering in 1950. For the next six years he specialized in research and development work at the Air Materiel Command, Wright Patterson AFB, Ohio. In June 1956 he returned to England as Chief of Logistics for SAC's 7th Air Division where he participated in writing the international agreement with Great Britain on the use of the Thor IBM. He was assigned to the Air Research and Development Command in 1959 and for four years he was director of the Minuteman program. General Phillips was promoted to Vice Commander of the Ballistic Systems Division in August 1963, and in January 1964 he moved to Washington to become deputy director of the Apollo program. His appointment as Director of the Apollo program came in October of that year.

* * *
- more -

* * *

GEORGE H. HAGE was appointed Deputy Director, Apollo Program, in January 1968, and serves as "general manager" assisting the Program Director in the management of Apollo developmental activities. In addition he is the Apollo Mission Director.

Hage was born in Seattle Washington, Oct. 7, 1925, and received his bachelor's degree in electrical engineering from the University of Washington in 1947. He joined Boeing that year and held responsible positions associated with the Bomarc and Minuteman systems, culminating in responsibility for directing engineering functions to activate the Cape Kennedy Minuteman Assembly and test complex in 1962. He then took charge of Boeing's unmanned Lunar Reconnaissance efforts until being named Boeing's engineering manager for NASA's Lunar Orbiter Program in 1963.

Hage joined NASA as Deputy Associate Administrator for Space Science and Applications (Engineering) July 5, 1967, and was assigned to the Apollo Program in October 1967 as Deputy Director (Engineering).

* * *

CHESTER M. LEE, U.S. NAVY (RET.) was appointed Assistant Apollo Mission Director in August 1966. He was born in New Derry, Pa., in 1919. Lee graduated from the U.S. Naval Academy in 1941 with a BS degree in electrical engineering. In addition to normal sea assignments he served with the Directorate of Research and Engineering, Office of Secretary of Defense and the Navy Polaris missile program. Lee joined NASA in August 1965 and served as Chief of Plans, Missions Operations Directorate, OMSF, prior to his present position.

* * *

COL. THOMAS H. McMULLEN (USAF) has been Assistant Mission Director, Apollo Program, since March 1968. He was born July 4, 1929, in Dayton, Ohio. Colonel McMullen graduated from the U.S. Military Academy in 1951 with a BS degree. He also received an MS degree from the Air Force Institute of Technology in 1964. His Air Force assignments included: fighter pilot, 1951-1953; acceptance test pilot, 1953-1962; development engineer, Gemini launch vehicle program office, 1964-1966; and Air Force Liaison Officer, 25th Infantry Division, 1967. He served in the Korean and Viet Nam campaigns and was awarded several high military decorations.

- more -

* * *

GEORGE P. CHANDLER, JR., Apollo 11 Mission Engineer, Apollo Operations Directorate, OMSF, Hq., was born in Knoxville, Tenn. Sept. 6, 1935. He attended grammar and high schools in that city was graduated from the University of Tennessee with a B.S. degree in electrical engineering in 1957. He received an army ROTC commission and served on active duty 30 months in the Ordnance Corps as a missile maintenance engineer in Germany. From 1960 until 1965 he was associated with Philco Corp. in Germany and in Houston, Texas. He joined the NASA Office of Manned Space Flight in Washington in 1965 and served in the Gemini and Apollo Applications operations offices before assuming his present position in 1967. Chandler was the mission engineer for Apollo 9 and 10.

* * *

MAJOR GENERAL JAMES W. HUMPHREYS, JR., USAF Medical Corps, joined NASA as Director of Space Medicine, on June 1, 1967. He was born in Fredericksburg, Va., on May 28, 1915. Humphreys graduated from the Virginia Military Institute with a BS degree in chemical engineering in 1935 and from the Medical College of Virginia with an MD in 1939. He served as a medical battalion and group commander in the European Theater in World War II and later as military advisor to the Iranian Army. Humphreys was awarded a master of science in surgery from the Graduate school of the University of Colorado in 1951. Prior to his association with NASA, General Humphreys was Assistant Director, USAID Vietnam for Public Health on a two year tour of duty under special assignment by the Department of State.

Manned Spacecraft Center

ROBERT R. GILRUTH, 55, Director, NASA Manned Spacecraft Center. Born Nashwauk, Minn. Joined NACA Langley Memorial Aeronautical Laboratory in 1936 working in aircraft stability and Control. Organized Pilotless Aircraft Research Division for transonic and supersonic flight research, 1945; appointed Langley Laboratory assistant director, 1952; named to manage manned space flight program, later named Project Mercury, 1958; named director of NASA Manned Spacecraft Center, 1961. BS and MS in aeronautical engineering from University of Minnesota; holds numerous honorary doctorate degrees. Fellow of the Institute of Aerospace Sciences, American Rocket Club and the American Astronautical Society. Holder of numerous professional society, industry and government awards and honorary memberships.

* * *

- more -

* * *

George M. Low, 43, manager, Apollo Spacecraft Program.
Born Vienna, Austria. Married to former Mary R. McNamara.
Children: Mark S. 17, Diane E. 15, G. David 13, John M.
11, and Nancy A. 6. Joined NACA Lewis Research Center
1949 specializing in aerodynamic heating and boundary layer
research; assigned 1958 to NASA Headquarters as assistant
director for manned space flight programs, later becoming
Deputy Associate Administrator for Manned Space Flight;
named MSC Deputy Director 1964; named manager, Apollo Spacecraft Program in April 1967. BS and MS in aeronautical engineering from Renssalaer Polytechnic Institute, Troy, N.Y.

* * *

CHRISTOPHER C. KRAFT, JR., 45, MSC Director of Flight
Operations. Born Phoebus, Va. Married to Former Elizabeth
Anne Turnbull of Hampton, Va. Children: Gordon T. 17, and
Kristi-Anne 14. Joined NACA Langley Aeronautical Laboratory
in 1945 specializing in aircraft stability and control; became member of NASA Space Task Group in 1958 where he developed basic concepts of ground control and tracking of manned
spacecraft. Named MSC Director of Flight Operations in November 1963. BS in aeronautical engineering from Virginia
Polytechnic Institute, Blacksburg, Va. Awarded honorary
doctorates from Indiana Institute of Technology and Parks
College of St. Louis University.

* * *

KENNETH S. KLEINKNECHT, 50, Apollo Spacecraft Program
manager for command and service modules. Born Washington,
D.C. Married to former Patricia Jean Todd of Cleveland,
Ohio. Children: Linda Mae 19, Patricia Ann 17, and Frederick W. 14. Joined NACA Lewis Research Center 1942 in
aircraft flight test; transferred to NACA Flight Research
Center 1951 in design and development work in advanced
research aircraft; transferred to NASA Space Task Group
1959 as technical assistant to the director; named manager
of Project Mercury 1962 and on completion of Mercury, deputy
manager Gemini Program in 1963; named Apollo Spacecraft Program manager for command and service modules early 1967 after
Gemini Program completed. BS in mechanical engineering
Purdue University.

* * *

- more -

* * *

CARROLL H. BOLENDER, 49, Apollo Spacecraft Program manager for lunar module. Born Clarksville, Ohio. He and his wife, Virginia, have two children--Carol 22 and Robert 13. A USAF Brigadier general assigned to NASA, Bolender was named lunar module manager in July 1967 after serving as a mission director in the NASA Office of Manned Space Flight. Prior to joining NASA, he was a member of a studies group in the office of the USAF chief of staff and earlier had worked on USAF aircraft and guided missile systems projects. During World War II, he was a night fighter pilot in the North African and Mediterranean theaters. He holds a BS from Wilmington College, Ohio, and an MS from Ohio State University.

* * *

DONALD K. SLAYTON, 45, MSC Director of Flight Crew Operations. Born Sparta, Wis. Married to the former Marjorie Lunney of Los Angeles. They have a son, Kent 12. Selected in April 1959 as one of the seven original Mercury astronauts but was taken off flight status when a heart condition was discovered. He subsequently became MSC Director of Flight Crew Operations in November 1963 after resigning his commission as a USAF major. Slayton entered the Air Force in 1943 and flew 56 combat missions in Europe as a B-25 pilot, and later flew seven missions over Japan. Leaving the service in 1946, he earned his BS in aeronautical engineering from University of Minnesota. He was recalled to active duty in 1951 as a fighter pilot, and later attended the USAF Test Pilot School at Edwards AFB, Calif. He was a test pilot at Edwards from 1956 until his selection as a Mercury astronaut. He has logged more than 4000 hours flying time---more than half of which are in jet aircraft.

* * *

CLIFFORD E. CHARLESWORTH, 37, Apollo 11 prime flight director (green team). Born Redwing, Minn. Married to former Jewell Davis, of Mount Olive, Miss. Children: David Alan 8, Leslie Anne 6. Joined NASA Manned Spacecraft Center April 1962. BS in physics from Mississippi College 1958. Engineer with Naval Mine Defense Lab, Panama City, Fla. 1958-60; engineer with Naval Ordnance Lab, Corona, Calif., 1960-61; engineer with Army Ordnance Missile Command, Cape Canaveral, Fla., 1961-62; flight systems test engineer, MSC Flight Control Division, 1962-65; head, Gemini Flight Dynamics Section, FCD, 1965-66; assistant Flight Dynamics Branch chief, FCD, 1966-68.

* * *

* * *

EUGENE F. KRANZ, 35, Apollo 11 flight director (white team) and MSC Flight Control Division chief. Born Toledo, Ohio. Married to former Marta I. Cadena of Eagle Pass, Texas, Children: Carmen 11, Lucy 9, Joan 7, Mark 6, Brigid 5 and Jean 3. Joined NASA Space Task Group October 1960. Supervisor of missile flight test for McDonnell Aircraft 1958-1960. USAF fighter pilot 1955-1958. McDonnell Aircraft flight test engineer 1954-55. BS in aeronautical engineering from Parks College, St. Louis University, 1954. Assigned as flight director of Gemini 3,4,7/6, 8, 9 and 12; Apollo 5, 8 and 9.

* * *

GLYNN S. LUNNEY, 32, Apollo 11 flight director (black team). Born Old Forge, Pa. Married to former Marilyn Jean Kurtz of Cleveland, Ohio. Children: Jenifer 8, Glynn 6, Shawn 5 and Bryan 3. Joined NACA Lewis Research Center August 1955 as college co-op employee. Transferred to NASA Space Task Group June 1959. Assigned as flight director of Gemini 9, 10, 11 and 12, Apollo 201,4, 7, 8 and 10. BS in aeronautical engineering from University of Detroit.

* * *

MILTON L. WINDLER, 37, Apollo 11 flight director (Maroon team). Born Hampton, Va. Married to former Betty Selby of Sherman, Texas. Children: Peter 12, Marion 9 and Cary 7. Joined NACA Langley Research Center June 1954. USAF fighter pilot 1955-58; rejoined NASA Space Task Group December 1959 and assigned to Recovery Branch of Flight Operations Division in development of Project Mercury recovery equipment and techniques. Later became chief of Landing and Recovery Division Operational Test Branch. Named Apollo flight director team April 1968.

* * *

CHARLES M. DUKE, 33, astronaut and Apollo 11 spacecraft communicator (CapCom). Born Charlotte, N. C. Married to former Dorothy M. Claiborne of Atlanta, Ga. Children: Charles M. 4, Thomas C. 2. Selected as astronaut in April 1966. Has rank of major in USAF, and is graduate of the USAF Aerospace Research Pilot School. Commissioned in 1957 and after completion of flight training, spent three years as fighter pilot at Ramstein Air Base, Germany. BS in naval sciences from US Naval Academy 1957; MS in aeronautics and astronautics from Massachusetts Institute of Technology 1964. Has more than 24 hours flying time, most of which is jet time.

* * *

* * *

RONALD E. EVANS, 35, astronaut and Apollo 11 spacecraft communicator (CapCom). Born St. Francis, Kans. Married to former Janet M. Pollom of Topeka, Kans. Children: Jaime D 9 and Jon P. 7. Selected an astronaut in April 1966. Has rank of lieutenant commander in U.S. Navy. Was flying combat missions from USS Ticonderoga off Viet Nam when selected for the astronaut program. Combat flight instructor 1961-1962; made two West Pacific aircraft carrier cruises prior to instructor assignment. Commissioned 1957 through University of Kansas Navy ROTC program. Has more than 3000 hours flying time, most of which is in jets. BS in electrical engineering from University of Kansas 1956; MS in aeronautical engineering from US Naval Postgraduate School 1964.

* * *

BRUCE McCANDLESS II, 32, astronaut and Apollo 11 spacecraft communicator (CapCom). Born Boston, Mass. Married to former Bernice Doyle of Rahway, N.J. Children: Bruce III 7 and Tracy 6. Selected as astronaut April 1966. Holds rank of lieutenant commander in US Navy. After flight training and earning naval aviator's wings in 1960, he saw sea duty aboard the carriers USS Forrestal and USS Enterprise, and later was assigned as instrument flight instructor at Oceana, Va. Naval Air Station. He has logged almost 2000 hours flying time, most of which is in jets. BS in naval sciences from US Naval Academy (second in class of 899) 1958; MS in electrical engineering from Stanford University 1965; working on PhD in electrical engineering at Stanford.

* * *

CHARLES A. BERRY, MD, 45, MSC Director of Medical Research and Operations. Born Rogers, Ark. Married to former Adella Nance of Thermal, Calif. Children: Mike, Charlene and Janice. Joined NASA Manned Spacecraft Center July 1962 as chief of Center Medical Operations Office; appointed MSC Director of Medical Research and Operations May 1966. Previously was chief of flight medicine in the office of the USAF Surgeon General 1959-62; assistant chief, then chief of department of aviation medicine at the School of Aviation Medicine, Randolph AFB, Texas 1956-59 and served as Project Mercury aeromedical monitor; Harvard School of Public Health aviation medicine residency 1955-56; base flight surgeon and command surgeon in stateside, Canal Zone and Carribean Assignments, 1951-1955. Prior to entering the USAF in 1951, Berry interned at University of California; service at San Francisco City and County Hospital and was for three years in general practice in Indio and Coachella, Calif. BA from University of California at Berkeley 1945; MD University of California Medical School, San Francisco, 1947; Master of public health, Harvard School of Public Health, 1956.

- more -

* * *

DR. WILMOT N. HESS, 42, MSC Director of Science and Applications. Born Oberlin, Ohio. Married to former Winifred Lowdermilk. Children: Walter C. 12, Alison L. 11 and Carl E. 9. Joined NASA Goddard Space Flight Center 1961 as chief of Laboratory for Theoretical Studies; transferred to NASA Manned Spacecraft Center 1967 as Director of Science and Applications. Previously leader, Plowshare Division of University of California Lawrence Radiation Laboratory 1959-61; physics instructor Oberlin College 1948-1949; physics instructor Mohawk College 1947. BS in electrical engineering from Columbia University 1946; MA in physics from Oberlin College 1949; and PhD in physics from University of California 1954.

* * *

DR. P. R. BELL, 56, chief MSC Lunar and Earth Sciences Division and manager of Lunar Receiving Laboratory. Born Fort Wayne, Indiana. Married to the former Mozelle Rankin. One son, Raymond Thomas 27. Joined NASA Manned Spacecraft Center July 1967. Formerly with Oak Ridge National Laboratories in thermonuclear research, instrumentation and plasma physics, 1946-67; MIT Radiation Laboratories in radar systems development, 1941-46; National Defense Research Committee Project Chicago, 1940-41. Holds 14 patents on electronic measurement devices, thermonuclear reactor components. BS in chemistry and doctor of science from Howard College, Birmingham, Ala.

* * *

JOHN E. McLEAISH, 39, Apollo 11 mission commentator and chief, MSC Public Information Office. Born Houston, Texas. Married to former Patsy Jo Thomas of Holliday, Texas. Children: Joe D. 19, Carol Ann 14, John E. Jr. 14. Joined NASA Manned Spacecraft Center 1962, named Public Information Office chief July 1968. Prior to joining NASA McLeaish was a USAF information officer and rated navigator from 1952 to 1962. BA in journalism from University of Houston. Assigned to mission commentary on Gemini 11 and 12 and Apollo 6 and 8.

* * *

JOHN E. (JACK) RILEY, 44, Apollo 11 mission commentator and deputy chief MSC Public Information Office. Born Trenton, Mo. Married to former Patricia C. Pray of Kansas City, Kans. Children: Kevin M. 17, Sean P. 15, Kerry E. 13, Brian T. 9

and Colin D. 6. Joined NASA Manned Spacecraft Center Public Information Office April 1963. Assigned to mission commentary on Gemini 9, 10 and 11 and Apollo 7, 9 and 10. PIO liaison with Apollo Spacecraft Program Office. Prior to joining NASA was public relations representative with General Dynamics/Astronautics 1961-63; executive editor, Independence, Mo. Examiner 1959-61; city editor, Kansas City Kansan 1957-59; reporter, Cincinnati, Ohio Times-Star 1957; reporter-copy editor, Kansas City Kansan 1950-57. Served in US Navy in Pacific-Asiatic Theaters 1942-46. BA in journalism University of Kansas.

* * *

DOUGLAS K. WARD, 29, born Idaho Falls, Idaho. Married to former Susan Diane Sellery of Boulder, Colorado. Children: Edward 7; Elisabeth, 4; and Cristina, 4. Joined NASA Public Affairs Office June 1966. Responsible for news media activities related to engineering and development and administrative operations at MSC. Assigned to mission commentary on Apollo 7, 8, and 10. BA in political science from the University of Colorado. Before joining NASA worked for two years with the U. S. Information Agency, Voice of America, writing and editing news for broadcast to Latin America and served as assistant space and science editor for the VOA news division.

* * *

(ROBERT) TERRY WHITE, 41, born Denton, Texas. Married to former Mary Louise Gradel of Waco, Texas. Children: Robert Jr., 4, and Kathleen, 2. Joined NASA Manned Spacecraft Center Public Affairs Office April 1963. Was editor of MSC Roundup (house organ) for four years. Assigned to mission commentary on 12 previous Gemini and Apollo missions. BA in journalism from North Texas State University. Prior to joining NASA, was with Employers Casualty Company, Temco Aircraft Corporation, (now LTV), Johnston Printing Company and Ayres Compton Associates, all of Dallas, Texas.

Marshall Space Flight Center

DR. WERNHER VON BRAUN became the director of MSFC when it was created in 1960. As a field center of NASA, the Marshall Center provides space launch vehicles and payloads, conducts related research, and studies advanced space transportation systems. Dr. von Braun was born in Wirsitz, Germany, on March 23, 1912. He was awarded a bachelor of science degree at the age of 20 from the Berlin Institute of Technology.

* * *

- more -

Two years later, he received his doctorate in physics from
the University of Berlin. He was technical director of
Germany's rocket program at Peenemunde. Dr. von Braun
came to the U.S. in 1945, under a contract to the U.S.
Army, along with 120 of his Peenemunde colleagues. He
directed high altitude firings of V-2 rockets at White
Sands Missile Range, N.M. and later became the project
director of the Army's guided missile development unit in
Fort Bliss. In 1950 he was transferred to Redstone Arsenal,
Ala. The Redstone, the Jupiter and the Pershing missile
systems were developed by the von Braun team. Current programs include the Saturn IB and the Saturn V launch vehicles
for Project Apollo, the nation's manned lunar landing program
and participation in the Apollo Applications program.

* * *

DR. EBERHARD F.M.REES is deputy director, technical,
of NASA-Marshall Space Flight Center. Dr. Rees was born
April 28, 1908, in Trossingen, Germany. He received his
technical education in Stuttgart and at Dresden Institute
of Technology. He graduated from Dresden in 1934 with a
master of science degree in mechanical engineering. During
World War II. Dr. Rees worked at the German Guided Missile
Center in Peenemunde. He came to the United States in 1945
and worked in the Ordnance Research and Development, Sub-
Office (rocket), at Fort Bliss. In 1950 the Fort Bliss activities were moved to Redstone Arsenal, Ala. Rees, who became
an American citizen in 1954, was appointed deputy director
of Research and Development of Marshall Space Flight Center
in 1960. He held this position until his appointment in
1963 to deputy director, technical.

* * *

DR. HERMANN K. WEIDNER is the director of Science and
Engineering at the Marshall Space Flight Center. Dr. Weidner
has had long and varied experience in the field of rocketry.
He became a member of the Peenemunde rocket development group
in Germany in 1941. In 1945, he came to the United States as
a member of the von Braun research and development team.
During the years that followed, this group was stationed at
Fort Bliss, as part of the Ordnance Research and Development.
After the Fort Bliss group was transferred to Huntsville, Dr.
Weidner worked with the Army Ballistic Missile Agency at Redstone Arsenal. He was formerly deputy director of the Pro-
pulsion and Vehicle Engineering Laboratory. He was also
director of propulsion at MSFC. Dr. Weidner received his
U.S. citizenship in April of 1955.

- more -

* * *

MAJ. GEN. EDMUND F. O'CONNOR is director of Program
Management at NASA-Marshall Space Flight Center. He is
responsible for the technical and administrative management of Saturn launch vehicle programs and that portion
of the Saturn/Apollo Applications Program assigned to
Marshall. O'Connor was born on March 31, 1922 in Fitch-
burg, Mass. He graduated from West Point in 1943, he
has a bachelor of science in both military engineering
and in aeronautical engineering. During World War II,
O'Connor served in Italy with the 495th Bombardment
Group, and held several other military assignments around
the world. In 1962 he went to Norton Air Force Base, as
deputy director of the Ballistic Systems Division, Air
Force Systems Command, He remained in that position until
1964 when he became director of Industrial Operations (now
designated Program Management) at Marshall Space Flight
Center.

* * *

LEE B. JAMES is the manager of the Saturn Program Office
in Program Management, Marshall Space Flight Center. A retired
Army Colonel, he has been in the rocket field since its infancy.
He started in 1947 after graduating in one of the early classes
of the Army Air Defense School at Fort Bliss. He is also a
graduate of West Point and he holds a master's degree from the
University of Southern California at Los Angeles. He joined
the rocket development team headed by Dr. Wernher von Braun
in 1956. When the team was transferred from the Department of
Defense to the newly created NASA in 1960, James remained as
director of the Army's Research and Development Division at
Redstone Arsenal. In 1961-62, he was transferred to Korea for
a one year tour of duty. After the assignment in Korea, he
was transferred by the Army to NASA-MSFC. In 1963 he became
manager of the Saturn I and IB program. For a year he served
in NASA Headquarters as deputy to the Apollo Program manager.
He returned in 1968 to manage the Saturn V program.

* * *

MATTHEW W. URLAUB is manager of the S-IC stage in the
Saturn Program Office at NASA-Marshall Space Flight Center.
Born September 23, 1927 in Brooklyn, he is a graduate of
Duke University where he earned his bachelor of science degree in mechanical engineering. Urlaub entered the army
in 1950 and finished his tour of duty in 1955. During the
period of 1952-1953 he completed a one year course at the
Ordnance Guided Missile School at Redstone Arsenal. Upon
becoming a civilian he became a member of the Army Ballistic

Missile Agency's Industrial Division Staff at Redstone Arsenal.
Specifically, he was the ABMA senior resident engineer for the
Jupiter Program at Chrysler Corporation in Detroit. He transferred to MSFC in 1961. The field in which he specializes
is project engineering/management.

* * *

ROY E. GODFREY performs dual roles, one as deputy manager
of the Saturn program and he is also the S-II stage manager.
Born in Knoxville on November 23, 1922, he earned a bachelor
of science degree in mechanical engineering at the University of
Tennessee. Godfrey served as second lieutenant in the Air Force
during WWII and began his engineering career with TVA. In
1953 he was a member of the research and development team at
Redstone Arsenal, when he accepted a position with the Ordnance
Missile laboratories. When the Army Ballistic Missile Agency
was created in 1956 he was transferred to the new agency. He
came to Marshall Center in 1962 to become the deputy director
of the Quality and Reliability Assurance Laboratory.

* * *

JAMES C. McCULLOCH is the S-IVB stage project manager
in the Saturn Program Office at the NASA-Marshall Space Flight
Center. A native of Alabama, he was born in Huntsville on
February 27, 1920. McCulloch holds a bachelor of science degree in mechanical engineering from Auburn University, and a
master's degree in business administration from Xavier University. Prior to coming to the Marshall Center in 1961, he had
been associated with Consolidated - Vultee Aircraft Corp.,
National Advisory Committee for Aeronautics; Fairchild Engine
and Airplane Corp., and General Electric Co.

* * *

FREDERICH DUERR is the instrument unit manager in the
Saturn Program Office at NASA-Marshall Space Flight Center.
Born in Munich, Germany, on January 26, 1909, he is a graduate of Luitpold Oberealschule and the Institute of Technology,
both in Munich. He holds B.S. and M.S. degrees in electrical
engineering. Duerr specializes in the design of electrical
network systems for the rocket launch vehicles. Duerr joined
Dr. Wernher von Braun's research and development team in 1941
at Peenemuende, and came with the group to the U.S. in 1945.
This group, stationed at White Sands, N.M., was transferred
to Huntsville in 1950 to form the Guided Missile Development
Division of the Ordnance Missile Laboratories at Redstone
Arsenal.

* * *

- more -

DR. FRIDTJOF A. SPEER is manager of the Mission Operations Office in Program Management at the NASA-Marshall Space Flight Center. A member of the rocket research and development team in Huntsville since March 1955, Dr. Speer was assistant professor at the Technical University of Berlin and Physics Editor of the Central Chemical Abstract Magazine in Berlin prior to coming to this country. He earned both his master's degree and Ph.D. in physics from the Technical University. From 1943 until the end of the war, he was a member of the Guided Missile Development group at Peenemunde. Dr. Speer was chief of the Flight Evaluation and Operations Studies Division prior to accepting his present position in August 1965. He became a U.S. citizen in 1960.

* * *

WILLIAM D. BROWN is manager, Engine Program Office in Program Management at MSFC. A native of Alabama, he was born in Huntsville on December 17, 1926. He is a graduate of Joe Bradley High School in Huntsville and attended Athens College and Alabama Polytechnic Institute to earn his bachelor of science degree in chemical engineering. Following graduation from Auburn University in 1951, he returned to Huntsville to accept a position with the Army research and development team at Redstone Arsenal, where he was involved in catalyst development for the Redstone missile. Shortly after the Army Ballistic Missile Agency was activated at Redstone, Brown became a rocket power plant engineer with ABMA. He transferred enmasse to the Marshall Space Flight Center when that organization was established in 1960.

* * *

Kennedy Space Center

DR. KURT H. DEBUS, Director, Kennedy Space Center, has
been responsible for many state of the art advances made in
launch technology and is the conceptual architect of the
Kennedy Space Center with its mobile facilities suitable for
handling extremely large rockets such as the Saturn V. Born
in Frankfurt, Germany, in 1908, he attended Darmstadt University
where he earned his initial and advanced degrees in mechanical
engineering. In 1939, he obtained his engineering doctorate and
was appointed assistant professor at the University. During this
period he became engaged in the rocket research program at
Peenemunde. Dr. Debus came to the United States in 1945 and
played an active role in the U.S. Army's ballistic missile
development program. In 1960, he was appointed Director of the
Launch Operations Directorate, George C. Marshall Space Flight
Center, NASA, at Cape Canaveral. He was appointed to his
present post in 1962. He brought into being the government/in-
dustry launch force which has carried out more than 150 successful
launches, including those of Explorer I, the Free World's first
satellite, the first manned launch and the Apollo 8 flight, first
manned orbit of the moon.

* * *

MILES ROSS, Deputy Director, Center Operations, Kennedy
Space Center, is responsible for operations related to engineer-
ing matters and the conduct of the Center's technical operations.
He has held the position since September 1967. Born in Brunswick,
N.J., in 1919, he is a graduate of Massachusetts Institute of
Technology where he majored in Mechanical Engineering and
Engineering Administration. Prior to his assignment at the
Kennedy Space Center, Ross was a project manager of the Air Force
Thor and Minuteman Missile systems with TRW, Inc. He was later
appointed Director of Flight Operations and Manager of Florida
Operations for TRW.

* * *

ROCCO A. PETRONE, Director of Launch Operations, Kennedy
Space Center, is responsible for the management and technical
direction of preflight operations and integration, test, check-
out and launch of all space vehicles, both manned and unmanned.
Born in Amsterdam, N.Y., in 1926, he is a 1946 graduate of the
U.S. Military Academy and received a Masters Degree in
Mechanical Engineering from Massachusetts Institute of Technology
in 1951. His career in rocketry began shortly after graduation
from MIT when he was assigned to the Army's Redstone Arsenal,
Huntsville, Ala. He participated in the development of the
Redstone missile in the early 1950's and was detailed to the
Army's General Staff at the Pentagon from 1956 to 1960. He
came to KSC as Saturn Project Officer in 1960. He later became
Apollo Program Manager and was appointed to his present post
in 1966.

* * *

* * *

RAYMOND L. CLARK, Director of Technical Support, Kennedy Space Center, is responsible for the management and technical direction of the operation and maintenance of KSC's test and launch complex facilities, ground support equipment and ground instrumentation required to support the assembly, test, checkout and launch of all space vehicles - both manned and unmanned. Born in Sentinel, Oklahoma, in 1924, Clark attended Oklahoma State University and is a 1945 graduate of the U.S. Military Academy with a degree in military science and engineering. He received a master of science degree in aeronautics and guided missiles from the University of Southern California in 1950 and was a senior project officer for the Redstone and Jupiter missile projects at Patrick AFB from 1954 to 1957. He joined KSC in 1960. Clark retired from the Army with the rank of lieutenant colonel in 1965.

* * *

G. MERRITT PRESTON, Director of Design Engineering, Kennedy Space Center, is responsible for design of ground support equipment, structures and facilities for launch operations and support elements at the nation's Spaceport. Born in Athens, Ohio, in 1916, he was graduated from Rensselaer Polytechnic Institute in New York with a degree in aeronautical engineering in 1939. He then joined the National Advisory Committee for Aeronautics (NACA) at Langley Research Center, Virginia, and was transferred in 1942 to the Lewis Flight Propulsion Center at Cleveland, Ohio, where he became chief of flight research engineering in 1945. NACA's responsibilities were later absorbed by NASA and Preston played a major role in Project Mercury and Gemini manned space flights before being advanced to his present post in 1967.

* * *

FREDERIC H. MILLER, Director of Installation Support, Kennedy Space Center, is responsible for the general operation and maintenance of the nation's Spaceport. Born in Toledo, Ohio, in 1911, he claims Indiana as his home state. He was graduated from Purdue University with a bachelor's degree in electrical engineering in 1932 and a master's degree in business administration from the University of Pennsylvania in 1949. He is a graduate of the Industrial College of the Armed Forces and has taken advanced management studies at the Harvard Business School. He entered the Army Air Corps in 1932, took his flight training at Randolph and Kelly Fields, Texas, and held various ranks and positions in the military service before retiring in 1966 as an Air Force major general. He has held his present post since 1967.

* * *

* * *

REAR ADMIRAL RODERICK O. MIDDLETON, USN, is Apollo Program
Manager, Kennedy Space Center, a post he has held since August,
1967. Born in Pomona, Fla., in 1919, he attended Florida Southern
College in Lakeland and was graduated from the U.S. Naval Academy
in 1937. He served in the South Pacific during World War II and
was awarded a master of science degree from Harvard University in
1946. He joined the Polaris development program as head of the
Missile Branch in the Navy's Special Project Office in Washington,
D.C., and was awarded the Legion of Merit for his role in the
Polaris project in 1961. He held a number of command posts, in-
cluding that as Commanding Officer of the USS Observation Island,
Polaris missile test ship, before being assigned to NASA in
October 1965.

* * *

WALTER J. KAPRYAN, Deputy Director of Launch Operations,
Kennedy Space Center, was born in Flint, Michigan, in 1920. He
attended Wayne University in Detroit prior to entering the Air
Force as a First Lieutenant in 1943. Kapryan joined the Langley
Research Center, National Advisory Committee for Aeronautics
(NACA) in 1947 and the NASA Space Task Group at Langley in March,
1959. He was appointed project engineer for the Mercury Redstone 1
spacecraft and came to the Cape in 1960 with that spacecraft. In
1963, he established and headed the Manned Spacecraft Center's
Gemini Program Office at KSC, participating in all 10 manned Gemini
flights as well as Apollo Saturn 1B and Saturn V missions before
advancement to his present post.

* * *

DR. HANS F. GRUENE, Director, Launch Vehicle Operations,
Kennedy Space Center, is responsible for the preflight testing,
preparations and launch of Saturn vehicles and operation and
maintenance of associated ground support systems. Born in
Braunschweig, Germany, in 1910, he earned his degrees in
electrical engineering at the Technical University in his hometown. He received his PhD in 1941 and began his career in guided
missile work as a research engineer at the Peenemunde Guided
Missile Center in 1943. He came to the United States with the
Army's Ordnance Research and Development Facility at Fort Bliss,
Texas, in 1945 and held a number of management posts at the Marshall
Space Flight Center, Huntsville, Ala., before being permanently
assigned to NASA's Florida launch site in June 1965.

* * *

* * *

JOHN J. WILLIAMS, Director, Spacecraft Operations, Kennedy Space Center, is responsible to the Director of Launch Operations for the management and technical integration of KSC Operations related to preparation, checkout and flight readiness of manned spacecraft. Born in New Orleans, La., in 1927, Williams was graduated from Louisiana State University with a bachelor of science degree in electrical engineering in 1949. Williams performed engineering assignments at Wright Patterson Air Force Base, Dayton, Ohio, and the Air Force Missile Test Center, Patrick AFB, Florida, before joining NASA in 1959. Williams played important roles in the manned Mercury and Gemini programs before moving to his current post in 1964.

* * *

PAUL C. DONNELLY, Launch Operations Manager, Kennedy Space Center, is responsible for the checkout of all manned space vehicles, including both launch vehicle and spacecraft. Born in Altoona, Pa., in 1923, Donnelly attended Grove City College in Pennsylvania, the University of Virginia and the U.S. Navy's electronics and guided missile technical schools. Donnelly performed engineering assignments at naval facilities at Chincoteague, Va., and Patuxent Naval Air Station, Md. Prior to assuming his present post, he was Chief Test Conductor for manned spacecraft at Cape Kennedy for the Manned Spacecraft Center's Florida Operations, his responsibilities extending to planning, scheduling and directing all manned spacecraft launch and prelaunch acceptance tests.

* * *

ROBERT E. MOSER, Chief, Test Planning Office, Launch Operations Directorate, Kennedy Space Center, is responsible for developing and coordinating KSC launch operations and test plans for the Apollo/Saturn programs. Born in Johnstown, Pa., in 1928, Moser regards Daytona Beach, Fla., as his hometown. A 1950 graduate of Vanderbilt University with a degree in electrical engineering, Moser has been associated with the U.S. space program since 1953 and served as test conductor for the launches of Explorer 1, the first American satellite; Pioneer, the first lunar probe; and the first American manned space flight - Freedom 7 - with Astronaut Alan B. Shepard aboard.

* * *

-more-

* * *

ISOM A. RIGELL, Deputy Director for Engineering, Launch Vehicle Operations, Kennedy Space Center, is responsible for all Saturn V launch vehicle engineering personnel in the firing room during prelaunch preparations and countdown, providing on-site resolution for engineering problems. Born in Slocomb, Ala., in 1923, Rigell is a 1950 graduate of the Georgia Institute of Technology with a degree in electrical engineering. He has played an active role in the nation's space programs since May, 1951.

* * *

ANDREW J. PICKETT, Chief, Test and Operations Management Office, Directorate of Launch Vehicle Operations, Kennedy Space Center, is responsible for directing the overall planning of Saturn launch vehicle preparation and prelaunch testing and checkout. Born in Shelby County, Ala., Pickett is a 1950 graduate of the University of Alabama with a degree in mechanical engineering. A veteran of well over 100 launches, Pickett began his rocketry career at Huntsville, Ala., in the early 1950s. He was a member of the Army Ballistic Missile Agency launch group that was transferred to NASA in 1960.

* * *

GEORGE F. PAGE, Chief of the Spacecraft Operations Division, Directorate of Launch Operations, Kennedy Space Center, is responsible for pre-flight checkout operations, countdown and launch of the Apollo spacecraft. Prior to his present assignment, Page was Chief Spacecraft Test Conductor and responsible for prelaunch operations on Gemini and Apollo spacecraft at KSC. Born in Harrisburg, Pa., in 1924, Page is a 1952 graduate of Pennsylvania State University with a bachelor of science degree in aeronautical engineering.

* * *

* * *

GEORGE T. SASSEEN, Chief, Engineering Division, Spacecraft Operations, Kennedy Space Center, is responsible for test planning and test procedure definition for all spacecraft prelaunch operations. Born at New Rochelle, N.Y., in 1928, he regards Weston, Conn., as his home town. A 1949 graduate of Yale University with a degree in electrical engineering, he joined NASA in July 1961. Prior to his present appointment in 1967, he served as Chief, Ground Systems Division, Spacecraft Operations Directorate, KSC. Sasseen's spacecraft experience extends through the manned Mercury, Gemini and Apollo programs.

* * *

DONALD D. BUCHANAN, Launch Complex 39 Engineering Manager for the Kennedy Space Center Design Engineering Directorate, is responsible for continuing engineering support at Launch Complex 39. He played a key role in the design, fabrication and assembly of such complex mobile structures as the mobile launchers, mobile service structure and transporters. Born in Macon, Ga., in 1922, Buchanan regards Lynchburg, Va., as his home town. He is a 1949 graduate of the University of Virginia with a degree in mechanical engineering. During the Spaceport construction phase, he was Chief, Crawler-Launch Tower Systems Branch, at KSC.

Office of Tracking and Data Acquisition

GERALD M. TRUSZYNSKI, Associate Administrator for Tracking and Data Acquisition, has held his present position since January, 1968, when he was promoted from Deputy in the same office. Truszynski has been involved in tracking, communication, and data handling since 1947, at Edwards, Cal., where he helped develop tracking and instrumentation for the X-1, X-15, and other high speed research aircraft for the National Advisory Committee for Aeronautics (NACA), NASA's predecessor. He directed technical design and development of the 500-mile aerodynamic testing range at Edwards. Truszynski joined NACA Langley Laboratory in 1944, after graduation from Rutgers University with a degree in electrical engineering. He was transferred to Headquarters in 1960, and became Deputy Associate Administrator in 1961. He is a native of Jersey City, N.J.

* * *

-more-

* * *

H R BROCKETT was appointed Deputy Associate Administrator
for Tracking and Data Acquisition March 10, 1968, after serving
five years as director of operations. He began his career with
NACA, predecessor of NASA, in 1947, at the Langley Research Center,
Hampton, Va., in the instrumentation laboratory. In 1958-59 he
was a member of a group which formulated the tracking and ground
instrumentation plans for the United States' first round-the-world
tracking network for Project Mercury. In 1959, he was transferred
to NASA Headquarters as a technical assistant in tracking operations. Brockett was born Nov. 12, 1924, in Atlanta, Neb. He is
a graduate of Lafayette College, 1947.

* * *

NORMAN POZINSKY, Director of Network Support Implementation
Division, Office of Tracking and Data Acquisition, has been
associated with tracking development since 1959, when he was
detailed to NASA as a Marine Corps officer. He assisted in
negotiations for facilities in Nigeria, Canada, and other countries
for NASA's worldwide tracking network. He retired from the marines
in 1963, and remained in his present position. Before joining NASA
he was involved in rocket and guided missile development at White
Sands, N.M., and China Lake, Cal., and served in 1956-59 as assistant chief of staff, USMC for guided missile systems. Born in
New Orleans in 1917, he is a 1937 graduate of Tulane University,
the Senior Command and Staff College, and the U.S. Navy Nuclear
Weapons School.

* * *

FREDERICK B. BRYANT, Director of DOD Coordination Division,
Office of Tracking and Data Acquisition, has been involved with
technical problems of tracking since he joined NASA's Office of
Tracking and Data Acquisition as a staff scientist in 1960. He
headed range requirements planning until he assumed his present
position in September 1964. He has charge of filling technical
requirements for tracking ships and aircraft of the Department
of Defense in support of NASA flights. Before joining NASA,
Bryant spent 21 years as an electronic scientist at the U.S. Navy's
David Taylor Model Basin, Carderock, Md. He worked on instrumentation, guidance control, and test programs for ships, submarines,
and underwater devices. Bryant received a B.S. degree in electronic
engineering in 1937 at Virginia Polytechnic Institute.

* * *

-more-

* * *

CHARLES A. TAYLOR became Director of Operations, Communications, and ADP Division, OTDA, when he joined NASA in November 1968. He is responsible for management and direction of operations of OTDA facilities and has functional responsibility for all NASA Automatic Data Processing. He was employed from 1942 to 1955 at NASA Langley Research Center, Hampton, Va., in research instrumentation for high speed aircraft and rockets. He worked for the Burroughs Corp., Paoli, Pa., in 1955-62, and after that for General Electric Co., Valley Forge, Pa., where he had charge of reliability and quality assurance, and managed the NASA Voyager space probe program. Born in Georgia, April 28, 1919, he received a B.S. degree at Georgia Institute of Technology in 1942.

* * *

PAUL A. PRICE, Chief of Communications and Frequency Management, OTDA, has held his position since he joined the NASA Headquarters staff in 1960. He is responsible for long-range planning and programming stations, frequencies, equipment, and communications links in NASCOM, the worldwide communications network by which NASA supports its projects on the ground and in space flight. Before he came to NASA, Price was engaged in communications and electronics work for 19 years for the Army, Navy, and Department of Defense. Born January 27, 1913, in Pittsburgh, he received his education in the public schools and was graduated from the Pennsylvania State University in 1935, and did graduate work at the University of Pittsburgh.

* * *

JAMES C. BAVELY, Director of the Network Operations Branch, OTDA, is responsible for operations management of NASA's networks in tracking, communication, command, and data handling for earth satellites, manned spacecraft, and unmanned lunar and deep space probes. Bavely held technical positions in private industry, the Air Force and Navy before joining NASA in 1961, with extensive experience in instrumentation, computer systems, telemetry, and data handling. Born in Fairmont, W. Va. in 1924, he is a 1949 graduate of Fairmont State College, and has since completed graduate science and engineering courses at George Washington University, University of West Virginia and University of Maryland.

* * *

* * *

E.J. STOCKWELL, Program Manager of MSFN Operations, OTDA, has held his present position since April 1962, when he joined the NASA Headquarters staff. Before that he had charge of ground instrumentation at the Naval Air Test Center, Patuxent River, Md. Stockwell was born May 30, 1926 in Howell, Mich. He received his education at Uniontown, Pa., and Fairmont, W.Va. He attended Waynesburg College and earned a B.S. degree in science from Fairmont State College. He is a director of the International Foundation for the Advancement of Telemetry.

* * *

LORNE M. ROBINSON, MSFN Equipment Program Manager, OTDA, is responsible for new facilities and equipment supporting NASA's manned space flight projects. He has been with OTDA since July 1963. He joined NASA from the Space Division of North American Rockwell, Downey, Cal., where he had been a senior research engineer on manned flight projects for five years. Previously, he was engaged in research at the University of Michigan Research Institute and the Phillips Chemical Co., Dumas, Tex. He was born December 20, 1930, in Detroit. He holds a degree in chemical engineering from Carnegie Institute of Technology (1952) and electrical engineering from the University of Michigan (1958), and completed graduate courses at UCLA.

Goddard Space Flight Center

OZRO N. COVINGTON is the Assistant Director, Manned Flight Support at the Goddard Center. Before joining NASA in June 1961 he was with the U.S. Army Signal Missile Support Agency as Technical Director for fifteen years. He studied electrical engineering at North Texas Agricultural College in Arlington, Texas, before embarking on an extensive career in radar and communications applications and research and development.

* * *

* * *

HENRY F. THOMPSON is Deputy Assistant Director for Manned
Flight Support at the Goddard Center. Thompson graduated from
the University of Texas with a B.A. degree in 1949 and a B.S. in
1952. His studies included graduate work at Texas Western College
in ElPaso and at the New Mexico State College, Las Cruces. Before
joining NASA in 1959 he was Technical Director of the U.S. Army
Electronics Command at White Sands Missile Range, N.M.

* * *

LAVERNE R. STELTER is chief of the Communications Division
at the Goddard Center. Mr. Stelter received his B.S. in electrical
engineering from the University of Wisconsin in 1951. After working with the Army Signal Corps, he joined NASA in 1959 as head of
the Goddard Communications Engineering Section. In 1961 he was
appointed Ground Systems Manager for TIROS weather satellites and
was later assigned the same responsibility for Nimbus. He was
appointed to his present position in 1963.

* * *

H. WILLIAM WOOD is head of the Manned Flight Operations
Division at the Goddard Center. Before joining NASA he was a
group leader at the Langley Research Center with responsibility
for implementing the Project Mercury Network. Mr. Wood earned
his BSEE degree at the North Carolina State University.

Department of Defense

MAJOR GENERAL DAVID M. JONES is Commander, Air Force Eastern
Test Range and Department of Defense Manager for Manned Space
Flight Support Operations.

He was born December 18, 1913, at Marshfield, Oregon, and
attended the University of Arizona at Tucson from 1932 to 1936.
He enlisted in the Arizona National Guard and served one year in
the Cavalry prior to entering pilot training in the summer of
1937.

His military decorations include the Legion of Merit,
Distinguished Flying Cross with one Oak Leaf Cluster, Air Medal,
Purple Heart, Yum Hwei from the Chinese government, and the NASA
Exceptional Service Medal with one device.

* * *

* * *

REAR ADMIRAL FRED E. BAKUTIS is Commander, Task Force 130, the Pacific Manned Spacecraft Recovery Force, in addition to his duty assignment as Commander, Fleet Air Hawaii.

He was born November 4, 1912, in Brockton, Massachusetts, and graduated from the U.S. Naval Academy June 6, 1935.

Admiral Bakutis holds the Navy Cross, the Legion of Merit with Combat "V," the Distinguished Flying Cross with Gold Star and the Bronze Star Medal.

* * *

REAR ADMIRAL PHILIP S. McMANUS is the Navy Deputy to the Department of Defense Manager for Manned Space Flight Support Operations and Commander, Task Force 140, the Atlantic Manned Spacecraft Recovery Force.

He was born in Holyoke, Massachusetts, on July 18, 1919, and was commissioned an ensign in the U.S. Navy following his graduation from the U.S. Naval Academy, in 1942.

Admiral McManus' decorations include the Legion of Merit with combat "V;" Navy and Marine Corps Medal; Navy Commendation Medal with combat "V;" and two Bronze Stars. His campaign medals include the European-African-Middle Eastern Campaign Medal with three Bronze Campaign Stars and the Asiatic-Pacific Campaign Medal with one Silver and four Bronze Campaign Stars.

* * *

BRIGADIER GENERAL ALLISON C. BROOKS is the Commander of Aerospace Rescue and Recovery Service (ARRS). He has the major responsibilities for both planned and contingency air recovery operations during Project Apollo.

He was born in Pittsburgh, Pennsylvania, June 26, 1917. General Brooks attended high school in Pasadena, California, and earned his Bachelor of Science degree from the University of California, Berkeley, California, in 1938. He enlisted a year later as a flying cadet in the Air Force and was graduated from Kelly Field in 1940.

General Brooks was awarded the Legion of Merit with one Oak Leaf Cluster, the Distinguished Flying Cross with two Oak Leaf Clusters, the Soldier's Medal, the Bronze Star Medal, the Air Medal with seven Oak Leaf Clusters and the French Croix de Guerre.

* * *

-more-

* * *

COLONEL ROYCE G. OLSON is Director, Department of Defense Manned Space Flight Support Office, located at Patrick AFB, Florida. He was born March 24, 1917, and is a native of Illinois, where he attended the University of Illinois.

He is a graduate of the National War College and holder of the Legion of Merit and Air Medal, among other decorations.

Major Apollo/Saturn V Contractors

Contractor	Item
Bellcomm Washington, D.C.	Apollo Systems Engineering
The Boeing Co. Washington, D.C.	Technical Integration and Evaluation
General Electric-Apollo Support Dept., Daytona Beach, Fla.	Apollo Checkout, and Quality and Reliability
North American Rockwell Corp. Space Div., Downey, Calif.	Command and Service Modules
Grumman Aircraft Engineering Corp., Bethpage, N.Y.	Lunar Module
Massachusetts Institute of Technology, Cambridge, Mass.	Guidance & Navigation (Technical Management)
General Motors Corp., AC Electronics Div., Milwaukee, Wis.	Guidance & Navigation (Manufacturing)
TRW Inc. Systems Group Redondo Beach, Calif.	Trajectory Analysis LM Descent Engine LM Abort Guidance System
Avco Corp., Space Systems Div., Lowell, Mass.	Heat Shield Ablative Material
North American Rockwell Corp. Rocketdyne Div. Canoga Park, Calif.	J-2 Engines, F-1 Engines
The Boeing Co. New Orleans	First Stage (SIC) of Saturn V Launch Vehicles, Saturn V Systems Engineering and Inte- gration, Ground Support Equip- ment
North American Rockwell Corp. Space Div. Seal Beach, Calif.	Development and Production of Saturn V Second Stage (S-II)
McDonnell Douglas Astronautics Co. Huntington Beach, Calif.	Development and Production of Saturn V Third Stage (S-IVB)

International Business Machines Federal Systems Div. Huntsville, Ala.	Instrument Unit
Bendix Corp. Navigation and Control Div. Teterboro, N.J.	Guidance Components for Instrument Unit (Including ST-124M Stabilized Platform)
Federal Electric Corp.	Communications and Instrumentation Support, KSC
Bendix Field Engineering Corp.	Launch Operations/Complex Support, KSC
Catalytic-Dow	Facilities Engineering and Modifications, KSC
Hamilton Standard Division United Aircraft Corp. Windsor Locks, Conn.	Portable Life Support System; LM ECS
ILC Industries Dover, Del.	Space Suits
Radio Corp. of America Van Nuys, Calif.	110A Computer - Saturn Checkout
Sanders Associates Nashua, N.H.	Operational Display Systems Saturn
Brown Engineering Huntsville, Ala.	Discrete Controls
Reynolds, Smith and Hill Jacksonville, Fla.	Engineering Design of Mobile Launchers
Ingalls Iron Works Birmingham, Ala.	Mobile Launchers (ML) (structural work)
Smith/Ernst (Joint Venture) Tampa, Fla. Washington, D.C.	Electrical Mechanical Portion of MLs
Power Shovel, Inc. Marion, Ohio	Transporter
Hayes International Birmingham, Ala.	Mobile Launcher Service Arms
Bendix Aerospace Systems Ann Arbor, Mich	Early Apollo Scientific Experiments Package (EASEP)
Aerojet-Gen. Corp El Monte. Calif.	Service Propulsion System Engine

APOLLO 11
PRINCIPAL INVESTIGATORS AND INVESTIGATIONS
OF LUNAR SURFACE SAMPLES

Investigator	Institution	Investigation
Adams, J.B. Co-Investigator: Jones, R. L.	Caribbean Research Inst. St. Thomas, V.I. NASA Manned Spacecraft Center Houston, Texas	Visible and Near-Infrared Reflection spectroscopy of Returned Lunar Sample at CRI & Lunar Receiving Lab. (LRL)
Adler, I. Co-Investigators: Walter, L.S. Goldstein, J. I. Philpotts, J.A. Lowman, P.D. French, B. M.	NASA Goddard Space Flight Center, Greenbelt, Md.	Elemental Analysis by Electron Microprobe
Agrell, S.O. Co-Investigator: Muir, I.O.	University Cambridge, England	Broad Mineralogic Studies
Alvarez, L.W. Co-Investigators: Watt, R.D.	University of California, Berkeley, California	Search for Magnetic Monopoles at LRL
Anders, E. Co-Investigators: Keays, R. R. Ganapathy, R. Jeffery, P.M.	University of Chicago, Chicago	a) Determine 14 Elements By Neutron Activation Analysis b) Measure Cosmic Ray Induced Al26 Content

Investigator	Institution	Investigation
Anderson, O. Co-Investigators: Soga, N. Kumazawa, M.	Lamont Geol. Obs. Columbia Univ. Palisades, N.Y.	Measure Sonic Velocity, Thermal Expansivity, Specific Heat, Dielectric Constant, and Index of Refraction
Arnold, J.R. Co-Investigators: Suess, H.E. Bhandari, N. Shedlovsky, J. Honda, M. Lal, D.	Univ. Calif., San Diego La Jolla, Calif.	Determine Cosmic Ray and Solar Particle Activation Effects
Arrhenius, G.O. Co-Investigators: Reid, A. Fitzgerald, R.	Univ. Calif., San Diego La Jolla, Calif.	Determine Microstructure Characteristics and Composition
Barghoorn, E. Co-Investigator: Philpott, D.	Harvard Univ. Cambridge, Mass. NASA Ames Res. Center, Moffett Field, Calif.	Electron Microscopy of Returned Lunar Organic Samples
Bastin, J. Co-Investigator: Clegg, P.E.	Queen Mary College London, England	Measure Electric Properties and Thermal Conductivity
Bell, P.M. Co-Investigator: Finger, L.	Carnegie Institution of Washington, Washington D.C.	Determine Crystal Structure of Separated Mineral Phases
Biemann, K.	Mass. Inst. Tech. Cambridge, Mass.	Mass Spectrometric Analyses for Organic Matter in Lunar Crust

-more-

Investigator	Institution	Investigation
Birkebak, R.C. Co-Investigators: Cremers, C.J. Dawson, J.P.	Univ. Kentucky Lexington, Ky.	Measure Thermal Radiative Features and Thermal Conductivity
Bowie, S.H.U. Co-Investigators: Horne, J.E.T. Snelling, N.J.	Inst. of Geol. Sciences, London England	Determinative Mineralogy for Opaque Materials by Electron Microprobe, Distribution of Radioactive Material by Auto-Radiograph, Analysis for Pb, U and Th Isotopes by Mass Spectrometry
Brown, G. M. Co-Investigators: Emeleus, C.H. Holland, J.G. Phillips, R.	Univ. Durham Durham, England	Petrologic Analysis by Standard Methods; Electron Probe Analysis Reflected Light Microscopy
Burlingame, A.L. Co-Investigator: Biemann, K.	Univ. of Calif., Berkeley, Calif. Mass. Inst. Tech. Cambridge, Mass.	Organic Mass Spectrometer Development for LRL
Calvin, M. Co-Investigators: Burlingame, A.L.	Univ. of Calif., Berkeley, Calif.	Study of Lunar Samples by Mass Spectrometry (Computerized) and Other Analytical Instrumentation
Cameron, E.N.	Univ. Wisconsin Madison, Wis.	Determine Structure, Composition Texture, and Phases of Opaque Material by Many Methods

Investigator	Institution	Investigation
Carter, N.L.	Yale Univ. New Haven, Conn.	Determine Effects of Shock on Lunar Materials Using Optical X-Ray, and Electron Microscopic Methods
Chao, E.C.T. Co-Investigators: James, O.B. Wilcox, R.E. Minkin, J.A.	U.S. Geol. Survey Washington, D.C.	Shocked Mineral Studies by Optical, X-Ray and Microprobe Techniques
Clayton, R.N.	Univ. Chicago	Determine Stable Isotope of Oxygen
Cloud, P. Co-Investigator: Philpott, D.	Univ. Calif., Los Angeles NASA Ames Res. Ctr.	Electron Microscopy of Returned Lunar Organic Samples
Collett, L.S. Co-Investigator: Becker, A.	Geol. Survey, Canada	Determine Electrical Conductivity
Compston, W.C. Co-Investigators: Arriens, P.A. Chappell, B.W. Vernon, M.J.	Australian Nat. University, Canberra	Sr and Sr Isotopes By X-Ray Fluorescence and Mass Spectrometry
Dalrymple, G.B. Co-Investigator: Doell, R.R.	U.S. Geol. Survey, Menlo Park, Calif.	Measure Natural & Induced Thermolum inescence to determine History and Environmental Features of Lunar Materials
Davis, R. Co-Investigator: Stoenner, R.W.	Brookhaven Nat. Lab., L.I., New York	Determine Ar^{37}, Ar^{39} Content

Investigator	Institution	Investigation
Doell, R. R. Co-Investigators: Gromme, C.S. Senfle, F.	U.S. Geol. Survey Menlo Park, Calif.	Measurement of Magnetic Properties at LRL and USGS Laboratories, Survey of Remnant Magnetism of Lunar Samples in Vacuum in the LRL
Douglas, J.A.V. Co-Investigators: Currie, K.L. Dence, M.R. Traill, R.J.	Geol. Survey of Canada Ottawa, Canada	Petrologic, Mineralogic and Textural Studies
Duke, M.B. Co-Investigator: Smith, R.L.	U.S. Geol. Survey, Washington, D.C.	Determine Size Frequency Distribution, Physical Properties and Composition of Lunar Materials of Sub-100 Micron Grain Size
Edgington, J.A. Co-Investigator: Blair, I.M.	Queen Mary College, Univ. London Atomic Energy Res. Establishment	Measure Luminescent and Thermo-luminescent Properties Under Proton (147 MEV) Bombardment
Eglinton, G. Co-Investigator: Lovelock, J.E.	Univ. Bristol Bristol, England	To Establish the Precise Nature of Organic Compounds in Lunar Material
Ehmann, W.D. Co-Investigator: Morgan, J.W.	Univ. Kentucky Lexington, Ky.	Analysis for Major Rock Forming Elements using 14 MEV Neutron Activation
Engel, A.E. Co-Investigator: Engel, A.C.J.	Univ. Calif., San Diego La Jolla, Calif.	Wet Chemical Analysis for Major Elements
Epstein, S. Co-Investigator: Taylor, H.P.	Cal. Inst. Tech. Pasadena, Calif.	Determine Content of Stable Isotopes of O, C, H, and Si by Mass Spectrometry

Investigator	Institution	Investigation
Evans, H.T. Co-Investigators: Barton, P.B., Jr. Roseboom, E.H.	U.S. Geol. Survey, Washington, D.C.	Crystal Structures of Sulfides and Related Minerals
Fields, P.R. Co-Investigators: Hess, D.C. Stevens, C.	Argonne Nat. Lab. Argonne, Ill.	Measure by Mass Spectrometry the Isotopic Abundances of Heavy Elements
Fireman, E.L.	Smithsonian Inst. Astrophysical Obs. Cambridge, Mass.	(a) Measure the Ar^{37} and Ar^{39} Content by Mass Spectrometry (b) Determine Tritium Content by Low Level Counting Techniques
Fleischer, R.L. Co-Investigators: Hanneman, R.E. Kasper, J.S. Price, P.B. Walker, R.M.	General Electric Schnectady, N.Y. Washington Univ. St. Louis, Mo.	(a) Measure Structural Defects in Lunar Materials Through Study of Optical, Electrical and Mechanical Properties (b) Determine the Effect of Cosmic Radiation on Lunar Samples by Study of Fossil Tracks Resulting from Charged Particles
Fox, S. Co-Investigators: Harada, K. Mueller, G.	Univ. Miami Coral Gables, Fla.	Analysis of Organic Lunar Samples for ALPHA Amino Acids and Polymers Thereof
Fredriksson, K. Co-Investigator: Nelen, J.	Smithsonian Inst. Nat. Museum Washington, D.C.	Elemental Analysis by Electron Microprobe
Friedman, I. Co-Investigator: O'Neil, J.R.	U.S. Geol. Survey Denver, Colo.	Isotopic Composition of H, D, and Oxygen

Investigator	Institution	Investigation
Frondel, C. Co-Investigators: Klein, C. Ito, J.	Harvard Univ. Cambridge, Mass.	Broad Studies of the Texture, Composition, and Relationship of Minerals
Gast, P.W.	Lamont Geol. Obs. Columbia Univ., Palisades, N.Y.	Determine Concentration of the Alkali, Alkaline Earth and Lanthanide Elements by Mass Spectrometry
Gay, P. Co-Investigators: Brown, M.G. McKie, D.	Univ. Cambridge Cambridge, England	X-Ray Crystallographic Studies
Geake, J.E. Co-Investigator: Garlick, G.F.J.	Univ. Manchester Manchester, England	Measure Fluorescence Emission and other Excitation Spectra; Optical Polarization; X-Ray Fluorescence; Electron Spin Resonance; Neutron Activation Analysis
Geiss, J. Co-Investigators: Eberhardt, P. Grogler, N. Oeschger, H.	Univ. Berne Berne, Switzerland	Measure Rare Gas Content and Cosmic Ray Produced Tritium by Mass Spectrometry
Gold, T.	Cornell Univ. Ithaca, N.Y.	Particle Size Analysis, Photometric Studies of Radiation Effects from Several Types of Rays; Direct Measure of Radiative Properties; Dielectric Constant and Loss Tangent

Investigator	Institution	Investigation
Goles, G.G.	Univ. Oregon Eugene, Ore.	Elemental Abundances by Neutron Activation Analysis
Greenman, N.N. Co-Investigator: Cross, H.G.	McDonnell-Douglas Corp. Santa Monica, Calif.	Determine the Luminescence Spectra and Efficiencies of Lunar Material and Compare with Mineral Composition
Grossman, J.J. Co-Investigators: Ryan, J.A. Mukherjee, N.R.	McDonnell-Douglas Corp. Santa Monica, Calif.	Microphysical, Microchemical and Adhesive Characteristics of the Lunar Materials
Hafner, S. Co-Investigator: Virgo, D.	Univ. Chicago Chicago, Ill.	Using Mossbauer and NMR Techniques Measure the Oxidation State of Iron, Radiation Damage and Al, Na, Fe Energy State in Crystals
Halpern, B. Co-Investigator: Hodgson, G.W.	Stanford Univ. Palo Alto, Calif.	Determine Terrestrial & Extra-terrestrial Porphyrins in Association with Amino Acid Compounds
Hapke, B.W. Co-Investigators: Cohen, A.J. Cassidy, W.	Univ. Pittsburgh Pittsburgh, Pa.	Determine Effects of Solar Wind on Lunar Material
Haskin, L.A.	Univ. Wisconsin Madison, Wis.	Determine Rare Earth Element Content by Neutron Activation Analysis

Investigator	Institution	Investigation
Helsley, C.E. Co-Investigators: Burek, P.J. Oetking, P.	Grad. Res. Center of the Southwest Dallas, Texas	Remanent Magnetism Studies
Helz, A.W. Co-Investigator: Annell, C.S.	U.S. Geol. Survey Washington, D.C.	Special Trace Elements by Emission Spectroscopy
Herr, W. Co-Investigators: Kaufhold, J. Skerra, B. Herpers, U.	Univ. Cologne Cologne, Germany	a) Determine Mn^{53} Content by High Flux Neutron Bombardment b) Determine Age of Lunar Materials Using Fission Track Method c) Measure Thermoluminescence to Determine Effect of Intrinsic Radioactive and Cosmic Ray Particles and Thermal History
Herzenberg, C.L.	Ill. Inst. Of Tech. Chicago	Measure the Energy States of the Iron Bearing Minerals and Possible Effects of Cosmic Radiation
Hess, H.H. Co-Investigator: Otolara, G.	Princeton Univ. Princeton, N.J.	Determine Pyroxene Content by X-Ray and Optical Methods
Heymann, D. Co-Investigators: Adams, J.A.A. Fryer, G.E.	Rice Univ. Houston, Texas	Determine Rare Gases and Radio- active Isotopes by Mass Spectro- metry

-more-

Investigator	Institution	Investigation
Hintenberger, H. Co-Investigator: Begemann, R.	Max Planck Inst. Fur Chemie, Mainz, Germany	a) Abundance and Isotopic Composition of Hydrogen
Begemann, R. Schultz, L. Vilcsek, E. Wanke, H. Wlotzka, A.		b) Measure Concentration and Isotopic Composition of Rare Gases
Voshage, H. Wanke, H. Schultz, L.		c) Isotopic Composition of Nitrogen
Hoering, T. Co-Investigator: Kaplan, I.R.	Carnegie Inst., D.C. Univ. Calif., L.A.	Analytical Lunar Sample Analyses for C13/C12 and D/H of Organic Matter
Hurley, P.M. Co-Investigator: Pinson, W.H., Jr.	Mass. Inst. Tech. Cambridge, Mass.	Analyze for Rb, Sr, and their Isotopes
Jedwab, J.	Univ. Libre De Bruxelles Brussels, Belgium	Determine the Morphological, Optical and Petrographic Properties of Magnetite and its Chemical Composition by Electron Microprobe
Johnson, R.D.	NASA Ames Research Ctr.	Analysis of Lunar Sample for Organic Carbon Behind the Barrier System of the LRL

-more-

Investigator	Institution	Investigation
Kaplan, I.R. Co-Investigators: Berger, R. Schopf, J.W.	Univ. Calif., Los Angeles, Calif.	Ratios of Carbon Hydrogen, Oxygen, and Sulphur Isotope Ratios by Mass Spectrometry
Kanamori, H. Co-Investigators: Mizutani, H. Takeuchi, H.	Univ. Toyko Japan	Determine Elastic Constants by Shear/Compressions/Wave Velocity
Keil, K. Co-Investigators: Bunch, T.E. Prinz, M. Snetsinger, K.G.	Univ. New Mexico Albuquerque, N.M.	Elemental Analysis and Mineral Phase Studies by Electron Microprobe
King, E.A. Co-Investigators: Morrison, D.A. Greenwood, W.R.	NASA Manned Spacecraft Center	Non-Destructive Mineralogy & Petrology; Analysis of the Fine Size Fraction of Lunar Materials Including Vitreous Phases
Kohman, T.P. Co-Investigator: Tanner, J.T.	Carnegie Inst. of Tech. Pittsburgh, Pa.	Determine Isotopic Abundance of Pb, Sr, Os, Tl, Nd, and Ag by Mass Spectrometry
Kuno, H. Co-Investigator: Kushiro, I.	Univ. Tokyo Japan	Petrographic Analysis for Mineral Identification and Chemical Composition
Larochelle, A. Co-Investigator: Schwarz, E.J.	Geol. Survey, Ottawa, Canada	Thermomagnetic, Magnetic Susceptibility and Remanent Magnetism Studies

Investigator	Institution	Investigation
Lipsky, S.R. Co-Investigators: Horvath, C.G. McMurray, W.J.	Yale Univ. New Haven, Conn.	Identification of Organic Compounds in Lunar Material by Means of Gas Chromatography-Mass Spectrometry, NMR, High Speed Liquid Chromatography, and Variations on these techniques
Lovering, J.F. Co-Investigators: Butterfield, D. (a) Kleeman, J.D. (b) Veizer, J. (b) Ware, N.G. (c)	Australian Nat. Univ. Canberra, Australia	a) Neutron Activation for U, Th, K b) Fission Track Analysis for U c) Electron Microprobe Analysis for Elemental Composition
MacGregor, I.D. Co-Investigator: Carter, J.L.	Grad. Research Ctr. S.W. Dallas, Texas	Petrographic Analysis by X-Ray Diffraction and Optical Methods
Manatt, S.L. Co-Investigators: Elleman, D.D. Vaughan, R.W. Chan, S.I.	NASA Jet Propulsion Lab., Pasadena, Calif. Cal. Inst. Tech.	Nuclear Radio Frequency Analysis Including NMR & ESR Analysis for Oxygen, Hydrogen, Water Content and other Elements and their Chemical State
Mason, B. Co-Investigators: Jarosewich, E. Fredriksson, K. White, J.S.	Smithsonian Inst. Nat. Museum Washington, D.C.	Mineralogic Investigations
Maxwell, J.A. Co-Investigators: Abbey, S. Champ, W.H.	Geol. Survey, Canada Ottawa, Canada	Wet Chemical, X-Ray Fluorescence and Emission Spectroscopy; Flame Photometry for Major/Minor Elements: Atomic Absorption Spectroscopy

Investigator	Institution	Investigation
McKay, D.S. Co-Investigators: Anderson, D. H. Greenwood, W.R. Morrison, D.A.	Manned Spacecraft Center	Determine Morphology and Composition of Fine Particles Using Electron Microprobe and Scanning Electron Microscope. Analysis to be Undertaken After Quarantine and not within the LRL
Meinschein, W.G.	Indiana Univ. Bloomington, Ind.	Determine the Alkane C_{15} To C_{30} Content By Gas Chromatographic and Mass Spectrometric Techniques
Moore, C.	Arizona State Univ. Temple, Ariz.	Determine Total Concentration of Carbon and Nitrogen
Morrison, G.H.	Cornell Univ.	Elemental Analysis using Spark Source Mass Spectrometry
Muir, A.H., Jr.	North American Rockwell Corp. Science Center, Thousand Oaks, Calif.	Conduct Mossbauer Effect and Spectroscopic Study of Iron-Bearing Mineral Separates
Murthy, V.R.	Univ. Minnesota Minneapolis, Minn.	Determine Rare-Earth Elemental and Low Abundance Isotopes of K, Ca, V and Cr Content by Neutron Activation
Nagata, T. Co-Investigators: Ozima, M. Ishikawa, Y.	Univ. Tokyo Japan	Remanent Magnetism Studies
Nagy, B. Co-Investigator: Urey, H.	Univ. Calif., San Diego La Jolla, Calif.	The Presence or Absence of Lipids, Amino Acids, and "Polymer-Type" Organic Matter

Investigator	Institution	Investigation
Nash, D.B.	NASA Jet Propulsion Lab.	Measure Luminescence, and Physical/Chemical Reaction of Lunar Material to Bombardment by o.5 to 10 KeV Protons
O'Hara, M.J. Co-Investigator: Biggar, G.M.	Univ. Edinburgh Scotland	Hi-Pressure/Temperature Phase Studies, Determine Temperature of Crystallization of Minerals; Petrologic Studies
O'Kelley, G.D. Co-Investigators: Bell, P.R. Eldridge, J.S. Schonfeld, E. Richardson, K.A.	Oak Ridge Nat. Lab. Tennessee MSC ORNL MSC MSC	Develop the Equipment/Methods for LRL; Measure the K40, U, Th, and Cosmic Ray Induced Radionuclide Content
Oro, J. Co-Investigators: Zlatkis, A. Lovelock, J.E. Becker, R.S. Updegrove, W.S. Flory, D.A.	Univ. Houston Houston, Texas	A Comprehensive Study of the Carbonaceous and Organogenic Matter Present in Returned Lunar Samples with Combination of Gas Chromatographic and Mass Spectrometric Techniques
Oyama, V.I. Co-Investigators: Merek, E. Silverman, M.P.	Manned Spacecraft Ctr. NASA Ames Res. Ctr. Moffett Field, Calif.	Isolation and Culture of Viable Organisms
Peck, L.C.	U.S. Geol. Survey Denver, Colo.	Standard Wet Chemical Analytical Techniques for Major Elements
Pepin, R.O. Co-Investigator: Nier, A.O.C.	Univ. Minnesota Minneapolis, Minn.	Measure the Elemental and Isotopic Abundances of He, Ne, Ar, Kr, and Xe by Mass Spectrometry

Investigator	Institution	Investigation
Perkins, R.W. Co-Investigators: Wogman, N.A. Kaye, J.H. Cooper, J.A. Rancitelli, L.A.	Battelle Mem. Inst. Richland, Wash.	Non-Destructive Gamma-Ray Spectrometry for Cosmic Ray Induced and Natural Radio-Nuclides
Philpotts, J.A. Co-Investigators: Schnetzler, C. Masuda, A. Thomas, H.H.	NASA Goddard Space Flight Center, Greenbelt, Md.	Determine the Rare Earth Element Content Using Dilution Technique and Mass Spectrometry
Ponnamperuma, C.A. Co-Investigators: Oyama, V.I. Pollack, G. Gehrke, C.W. Zill, L.P.	NASA Ames Res. Center Moffett Field, Calif. Univ. Missouri Ames Res. Center	Analytical Lunar Sample Analyses for Amino Acids, Nucleic Acids, Sugars, Fatty Acids, Hydrocarbons, Porphyrins and Their Components
Quaide, W.L. Co-Investigators: Wrigley, R.C. (a) Debs, R.J. (a) Bunch, T.E. (b)	Ames Res. Center	a) By Non-Destructive Gamma-Ray Spectrometry Determine the Al26, Na22, and Mn54 Content b) Microscopic, X-Ray Diffraction Analysis to Determine the Effects of Shock on Minerals and Rocks
Ramdohr, P. Co-Investigator: ElGoresy, A.	Max Planck Institut Heidelberg, Germany	Identification of Opaque Minerals, Phases and Composition by X-Ray, Microprobe, and Microscopic Analysis

Investigator	Institution	Investigation
Reed, G.W. Co-Investigators: Huizenga, J. Jovanovic, S. Fuchs, L.	Argonne Nat. Lab. Argonne, Ill.	Concentration, Isotopic Composition and Distribution of Trace Elements by Neutron Activation Analysis
Reynolds, J.H. Co-Investigators: Rowe, M.W. Hohenberg, C.M.	Univ. Calif., Berkeley	(a) Rare Gas Content by Mass Spectrometry (b) Mass Spectrometry to Identify Cosmic Ray Produced Nuclides (c) Mass Spectrometry to Determine Rare Gas, K and U Content; Identify Cosmic Ray Produced Nuclides
Rho, J.H. Co-Investigators: Bauman, A.J. Bonner, J.F.	NASA Jet Propulsion Lab. Pasadena, California Cal. Inst. Tech.	Determine Metallic and Non-Metallic Porphyrin Content by Fluorescence Spectrophotometry
Richardson, K.A. Co-Investigators: McKay, D.S. Foss, T.H.	NASA Manned Spacecraft Center Houston, Texas	By Autoradiography and Alpha Particle Spectroscopy Identify Alpha Emitting Nuclides
Ringwood, A.E. Co-Investigator: Green, D.H.	Australian Nat'l Univ. Canberra	Petrographic Analysis by Study of Thin and Polished Sections
Robie, R.A.	U.S. Geol. Survey, D.C.	Calorimetry (Thermal Properties)
Roeder, E.	U.S. Geol. Survey, D.C.	Determine the Nature and Composition of Fluid Inclusions, if Present, in Lunar Material

Investigator	Institution	Investigation
Rose, H.W., Jr. Co-Investigators: Cuttitta, F. Dwornik, E.J.	U.S. Geol. Survey, D.C.	X-Ray Fluorescence Methods for Elemental Analysis
Ross, M. Co-Investigators: Warner, J. Papike, J.J. Clark, J.R.	U.S. Geol. Survey Washington, D.C. NASA MSC USGS, D.C.	Determine the Crystallographic Parameters and Composition of Pyroxenes, Micas, Amphiboles, and Host Silicate Minerals by X-Ray Diffraction and Electron Microprobe
Runcorn, S.K.	Univ. of Newcastle Upon Tyne, England	Magnetic Properties in Conjunction with Mineralogic Studies
Schaeffer, O.A. Co-Investigators: Zahringer, J. Bogard, D.	State Univ. of N.Y. at Stony Brook, N.Y. Max Planck Inst. - Germany Manned Spacecraft Ctr.	Determine Rare Gas Content by Mass Spectrometry at LRL
Schmidt, R.A. Co-Investigator: Loveland, W.D.	Oregon State Univ. Corvallis, Ore.	Determine Rare-Earth and Selected Trace Elements Content by Neutron Activation Analysis. Isotopes of Sm, Eu, Gd will be Determined by Mass Spectrometry
Schopf, W.	Univ. of Calif., L.A. Los Angeles	Micropaleontological Study Using Transmission and Scanning Electron Microscopy
Sclar, C.B. Co-Investigator: Melton, C.W.	Battelle Mem. Inst. Columbus, Ohio	Using Replication and Thin Section Electron Microscopy Determine the Damage in Minerals and Rocks Due to Shock
Scoon, J.H.	Univ. Cambridge England	Wet Chemical Analysis for Major Elements

-more-

Investigator	Institution	Investigation
Short, N.M.	Univ. Houston Houston, Texas	By Petrographic Studies Determine Effect of Shock on Rocks and Minerals and Predict Magnitude of Shock and Line of Impacting Missile
Silver, L.T. Co-Investigator: Patterson, C. C.	Calif. Inst. Tech. Pasadena, Calif.	Determine Lead Isotopes, Concentrations of U, Th, Pb, and their occurrence in Minerals
Simmons, G. Co-Investigators: Brace, W.F.)--Parts A & B only Wones, D.R.)	Mass. Inst. Tech. Cambridge, Mass.	a) Calculate Elastic Properties From Measurement of Compressional Shear Wave Velocities at STP b) Measure Thermal Conductivity, Expansion and Diffusivity at STP c) Determine Dielectric Constant, Resistivity
		Determine Thermal Properties at STP on Samples of Core From Lunar Surface
Sippel, R.F. Co-Investigator: Spencer, A.B.	Mobil Res. and Dev. Corp. Dallas, Texas	Apply Luminescence Petrography to Study of Lunar Materials
Skinner, B.J. Co-Investigator: Winchell, H.	Yale Univ. New Haven, Conn.	Examine the Returned Samples for Condensed Sublimates and if Present Determine the Mineral Phases Present and Elemental Composition
Smales, A.A.	Atomic Energy Research Estab., Harwell, England	Elemental and Isotopic Abundances by Neutron Activation Analysis and by Emission, Spark Source, and X-Ray Fluorescence Spectrography

Investigator	Institution	Investigation
Smith, J.V. Co-Investigators: Wyllie, P.J. Elders, W.A.	Univ. Chicago Chicago, Ill.	Mineralogic-Petrographic Analysis Using Microprobe, X-Ray Diffraction and Microscopic Methods
Stephens, D.R. Co-Investigator: Keeler, R.N.	Lawrence Radiation Lab. Livermore, California	Physical Properties-Equation of State
Stewart, D.B. Co-Investigators: Appleman, D.E. Papike, J.J. Clark, J.R. Ross, M.	U.S. Geol. Survey, D.C.	Crystal Structure and Stabilities of Feldspars
Strangway, D.W.	Univ. Toronto Canada	Determine Magnetic Properties Including Remanent, Susceptibility, Thermal, Demagnetization, Identify Magnetic Minerals
Tatsumoto, M. Co-Investigator: Doe, B.R.	U.S. Geol. Survey Denver, Colo.	Pb Analysis by Mass Spectrometry; U and Th Analysis by Mass Spectrometry and Alpha Spectrometry
Tolansky, S.	Royal Holloway College, Univ. London, England	Isotopic Abundances of U and Th by Microscopic Studies of Diamonds
Turekian, K.K.	Yale Univ. New Haven, Conn.	Determine 20 Elements Having Halflives Greater Than 3 Days By Neutron Activation Analysis

-more-

Investigator	Institution	Investigation
Turkevich, A.L.	Univ. of Chicago Chicago, Ill.	a) Determine Long Lived Isotopes of K, U, and Th by Gamma Ray Spectrometry b) Neutron Activation Analysis for U, Th, Bi, Pb, Tl, and Hg.
Turner, G.	Univ. Sheffield	Determine AR^{40}/Ar^{39} for Age Dating
Urey, H.C. Co-Investigator: <u>Marti, K.</u>	Univ. Calif., S.D. La Jolla, Calif.	Isotopic Abundances by Mass Spectroscopy
Von Engelhardt, W. Co-Investigators: <u>Stoffler, D.</u> Muller, W. Arndt, J.	Univ. Tubingen Tubingen, Germany	Petrographic Study to Determine Shock Effects
Walker, R.M.	Washington Univ. St. Louis, Mo.	a) Measure the Structural Damage to Crystalline Material by Several Techniques b) Geochronological Studies by Investigation of Fission Tracks From Radioactive and Cosmic Ray Particles
Wanke, H. Co-Investigators: <u>Begemann, F.</u> Vilcsek, E. Voshage, H.	Max Planck Inst. Fur Chemie, Mainz Germany	a) Determine K, Th, U Content
Begemann, F. Vilcsek, E.		b) Measure Cosmic Ray Induced Radioactive Nuclides C^{14} and Cl^{36}

This investigation continued on next page.

Investigator	Institution	Investigation
Rieder, R.		c) Major Elemental Abundances by Fast Neutron Activation
Rieder, R. Wlotzka, F.		d) Minor Elemental Abundances by Thermal Neutron Activation
Wanless, R.K. Co-Investigators: Stevens, R.D. Loveridge, W.D.	Geol. Survey, Ottawa, Canada	Determine Concentrations of Pb, U, Th, Rb, Sr, Ar, & K and the Isotopic Compositions of Pb and Sr.
Wasserburg, G.J. Co-Investigator: Burnett, D.S.	Cal. Inst. Tech.	Determine K, Ar, Rb, Sr and Rare Gas (He, Ne, Ar, Kr, Xe) Content by Mass Spectrometry
Wasson, J.T. Co-Investigator: Baedecker, P.A.	Univ. Calif., Los Angeles, Calif.	Elemental Abundances for Ga and Ge by Neutron Activation
Weeks, R.A. Co-Investigator: Kolopus, J.	Oak Rdige Nat. Lab. Oak Ridge, Tenn.	Determine the Valence State and Symmetry of the Crystalline Material Using Electron Spin and Nuclear Magnetic Resonance Techniques and Spin Lattice Relaxation Studies
Weill, D.F.	Univ. Oregon Eugene, Ore.	Determine Temperature of Rock Formation by Study of Plagioclase Properties
Wetherill, G.W.	Univ. Calif., Los Angeles, Calif.	Determine Isotopes of Rb, Sr, U, and Pb by Mass Spectrometry
Wiik, H.B. Co-Investigator: Ojanpera, P.M.	Geol. Survey Helsinki, Finland	Wet Chemical Methods to Determine Major Elemental Abundance

-more-

Investigator	Institution	Investigation
Wood, J.A. Co-Investigator: Marvin, U.B.	Smithsonian Inst. Astrophysical Obs. Cambridge, Mass.	Mineralogic and Petrologic Studies by Optical Microscopy, X-Ray Diffraction and Electron Microprobe Measurements
Zahringer, J. Co-Investigators: Kirsten, I. Lammerzahl, P.	Max Planck Inst. Heidelberg Heidelberg, Germany	By Microprobe Analysis and Mass Spectrometry Determine Gas Content and Solar Wind Particle Distribution
Zussman, J.	Univ. Manchester Manchester, England	Geochemical, Mineralogic, and Petrological Studies

APOLLO GLOSSARY

Ablating Materials--Special heat-dissipating materials on the surface of a spacecraft that vaporize during reentry.

Abort--The unscheduled termination of a mission prior to its completion.

Accelerometer--An instrument to sense accelerative forces and convert them into corresponding electrical quantities usually for controlling, measuring, indicating or recording purposes.

Adapter Skirt--A flange or extension of a stage or section that provides a ready means of fitting another stage or section to it.

Antipode--Point on surface of planet exactly 180 degrees opposite from reciprocal point on a line projected through center of body. In Apollo usage, antipode refers to a line from the center of the Moon through the center of the Earth and projected to the Earth surface on the opposite side. The antipode crosses the mid-Pacific recovery line along the 165th meridian of longitude once each 24 hours.

Apocynthion--Point at which object in lunar orbit is farthest from the lunar surface -- object having been launched from body other than Moon. (Cynthia, Roman goddess of Moon)

Apogee--The point at which a Moon or artificial satellite in its orbit is farthest from Earth.

Apolune--Point at which object launched from the Moon into lunar orbit is farthest from lunar surface, e.g.: ascent stage of lunar module after staging into lunar orbit following lunar landing.

Attitude--The position of an aerospace vehicle as determined by the inclination of its axes to some frame of reference; for Apollo, an inertial, space-fixed reference is used.

Burnout--The point when combustion ceases in a rocket engine.

Canard--A short, stubby wing-like element affixed to the launch escape tower to provide CM blunt end forward aerodynamic capture during an abort.

Celestial Guidance--The guidance of a vehicle by reference to celestial bodies.

-more-

Celestial Mechanics--The science that deals primarily with the effect of force as an agent in determining the orbital paths of celestial bodies.

Cislunar--Adjective referring to space between Earth and the Moon, or between Earth and Moon's orbit.

Closed Loop--Automatic control units linked together with a process to form an endless chain.

Deboost--A retrograde maneuver which lowers either perigee or apogee of an orbiting spacecraft. Not to be confused with deorbit.

Declination--Angular measurement of a body above or below celestial equator, measured north or south along the body's hour circle. Corresponds to Earth surface latitude.

Delta V--Velocity change.

Digital Computer--A computer in which quantities are represented numerically and which can be used to solve complex problems.

Down-Link--The part of a communication system that receives, processes and displays data from a spacecraft.

Entry Corridor--The final flight path of the spacecraft before and during Earth reentry.

Ephemeris--Orbital measurements (apogee, perigee, inclination, period, etc.) of one celestial body in relation to another at given times. In spaceflight, the orbital measurements of a spacecraft relative to the celestial body about which it orbited.

Escape Velocity--The speed a body must attain to overcome a gravitational field, such as that of Earth; the velocity of escape at the Earth's surface is 36,700 feet-per-second.

Explosive Bolts--Bolts destroyed or severed by a surrounding explosive charge which can be activated by an electrical impulse.

Fairing--A piece, part or structure having a smooth, stream-lined outline, used to cover a nonstreamlined object or to smooth a junction.

Flight Control System--A system that serves to maintain attitude stability and control during flight.

Fuel Cell--An electrochemical generator in which the chemical
    energy from the reaction of oxygen and a fuel is con-
    verted directly into electricity.

g or g Force--Force exerted upon an object by gravity or by
    reaction to acceleration or deceleration, as in a change
    of direction: one g is the measure of force required to
    accelerate a body at the rate of 32.16 feet-per-second.

Gimbaled Motor--A rocket motor mounted on gimbal; i.e.: on a
    contrivance having two mutually perpendicular axes of ro-
    tation, so as to obtain pitching and yawing correction moments.

Guidance System--A system which measures and evaluates flight
    information, correlates this with target data, converts
    the result into the conditions necessary to achieve the
    desired flight path, and communicates this data in the form
    of commands to the flight control system.

Heliocentric--Sun-centered orbit or other activity which has the
    Sun at its center.

Inertial Guidance--Guidance by means of the measurement and
    integration of acceleration from on board the spacecraft.
    A sophisticated automatic navigation system using gyro-
    scopic devices, accelerameters etc., for high-speed vehicles.
    It absorbs and interprets such data as speed, position, etc.,
    and automatically adjusts the vehicle to a pre-determined
    flight path. Essentially, it knows where it's going and
    where it is by knowing where it came from and how it got
    there. It does not give out any radio frequency signal so
    it cannot be detected by radar or jammed.

Injection--The process of boosting a spacecraft into a calcu-
    lated trajectory.

Insertion--The process of boosting a spacecraft into an orbit
    around the Earth or other celestial bodies.

Multiplexing--The simultaneous transmission of two or more sig-
    nals within a single channel. The three basic methods
    of multiplexing involve the separation of signals by time
    division, frequency division and phase division.

Optical Navigation--Navigation by sight, as opposed to inertial
    methods, using stars or other visible objects as reference.

Oxidizer--In a rocket propellant, a substance such as liquid
    oxygen or nitrogen tetroxide which supports combustion of
    the fuel.

-more-

Penumbra--Semi-dark portion of a shadow in which light is partly
cut off, e.g.: surface of Moon or Earth away from Sun where
the disc of the Sun is only partly obscured.

Pericynthion--Point nearest Moon of object in lunar orbit--object
having been launched from body other than Moon.

Perigee--Point at which a Moon or an artificial satellite in its
orbit is closest to the Earth.

Perilune--The point at which a satellite (e.g.: a spacecraft) in
its orbit is closest to the Moon. Differs from pericynthion
in that the orbit is Moon-originated.

Pitch--The movement of a space vehicle about an axis (Y) that is
perpendicular to its longitudinal axis.

Reentry--The return of a spacecraft that reenters the atmosphere
after flight above it.

Retrorocket--A rocket that gives thrust in a direction opposite
to the direction of the object's motion.

Right Ascension--Angular measurement of a body eastward along the
celestial equator from the vernal equinox (0 degrees RA) to
the hour circle of the body. Corresponds roughly to Earth
surface longitude, except as expressed in hrs:min:sec instead
of 180 degrees west and east from 0 degrees (24 hours=360
degrees).

Roll--The movements of a space vehicle about its longitudinal
(X) axis.

S-Band--A radio-frequency band of 1,550 to 5,200 megahertz.

Selenographic--Adjective relating to physical geography of Moon.
Specifically, positions on lunar surface as measured in
latitude from lunar equator and in longitude from a
reference lunar meridian.

Selenocentric--Adjective referring to orbit having Moon as center.
(Selene, Gr. Moon)

Sidereal--Adjective relating to measurement of time, position
or angle in relation to the celestial sphere and the vernal
equinox.

State vector--Ground-generated spacecraft position, velocity and
timing information uplinked to the spacecraft computer for
crew use as a navigational reference.

Telemetering--A system for taking measurements within an aerospace vehicle in flight and transmitting them by radio to a ground station.

Terminator--Separation line between lighted and dark portions of celestial body which is not self luminous.

Ullage--The volume in a closed tank or container that is not occupied by the stored liquid; the ratio of this volume to the total volume of the tank; also an acceleration to force propellants into the engine pump intake lines before ignition.

Umbra--Darkest part of a shadow in which light is completely absent, e.g.: surface of Moon or Earth away from Sun where the disc of the Sun is completely obscured.

Update pad--Information on spacecraft attitudes, thrust values, event times, navigational data, etc., voiced up to the crew in standard formats according to the purpose, e.g.: maneuver update, navigation check, landmark tracking, entry update, etc.

Up-Link Data--Information fed by radio signal from the ground to a spacecraft.

Yaw--Angular displacement of a space vehicle about its vertical (Z) axis.

-more-

APOLLO ACRONYMS AND ABBREVIATIONS

(Note: This list makes no attempt to include all Apollo
program acronyms and abbreviations, but several are listed
that will be encountered frequently in the Apollo 11 mission.
Where pronounced as words in air-to-ground transmissions,
acronyms are phonetically shown in parentheses. Otherwise,
abbreviations are sounded out by letter.)

AGS	(Aggs)	Abort Guidance System (LM)
AK		Apogee kick
APS	(Apps)	Ascent Propulsion System (LM)
		Auxiliary Propulsion System (S-IVB stage)
BMAG	(Bee-mag)	Body mounted attitude gyro
CDH		Constant delta height
CMC		Command Module Computer
COI		Contingency orbit insertion
CRS		Concentric rendezvous sequence
CSI		Concentric sequence initiate
DAP	(Dapp)	Digital autopilot
DEDA	(Dee-da)	Data Entry and Display Assembly (LM AGS)
DFI		Development flight instrumentation
DOI		Descent orbit insertion
DPS	(Dips)	Descent propulsion system
DSKY	(Diskey)	Display and keyboard
EPO		Earth Parking Orbit
FDAI		Flight director attitude indicator
FITH	(Fith)	Fire in the hole (LM ascent abort staging)
FTP		Full throttle position
HGA		High-gain antenna
IMU		Inertial measurement unit

IRIG	(Ear-ig)	Inertial rate integrating gyro
LOI		Lunar orbit insertion
LPO		Lunar parking orbit
MCC		Mission Control Center
MC&W		Master caution and warning
MSI		Moon sphere of influence
MTVC		Manual thrust vector control
NCC		Combined corrective maneuver
PDI		Powered descent initiation
PIPA	(Pippa)	Pulse integrating pendulous accelerometer
PLSS	(Pliss)	Portable life support system
PTC		Passive thermal control
PUGS	(Pugs)	Propellant utilization and gaging system
REFSMMAT	(Refsmat)	Reference to stable member matrix
RHC		Rotation hand controller
RTC		Real-time command
SCS		Stabilization and control system
SHE	(Shee)	Supercritical helium
SLA	(Slah)	Spacecraft LM adapter
SPS		Service propulsion system
TEI		Transearth injection
THC		Thrust hand controller
TIG	(Tigg)	Time at ignition
TLI		Translunar injection
TPF		Terminal phase finalization
TPI		Terminal phase initiate
TVC		Thrust vector control

CONVERSION FACTORS

	Multiply	By	To Obtain
Distance:			
	feet	0.3048	meters
	meters	3.281	feet
	kilometers	3281	feet
	kilometers	0.6214	statute miles
	statute miles	1.609	kilometers
	nautical miles	1.852	kilometers
	nautical miles	1.1508	statute miles
	statute miles	0.86898	nautical miles
	statute mile	1760	yards
Velocity:			
	feet/sec	0.3048	meters/sec
	meters/sec	3.281	feet/sec
	meters/sec	2.237	statute mph
	feet/sec	0.6818	statute miles/hr
	feet/sec	0.5925	nautical miles/hr
	statute miles/hr	1.609	km/hr
	nautical miles/hr (knots)	1.852	km/hr
	km/hr	0.6214	statute miles/hr
Liquid measure, weight:			
	gallons	3.785	liters
	liters	0.2642	gallons
	pounds	0.4536	kilograms
	kilograms	2.205	pounds

- more -

Multiply	By	To Obtain
Volume:		
cubic feet	0.02832	cubic meters
Pressure:		
pounds/sq inch	70.31	grams/sq cm

Propellant Weights

RP-1 (kerosene) --------- Approx. 6.7 pounds per gallon

Liquid Oxygen ----------- Approx. 9.5 pounds per gallon

Liquid Hydrogen --------- Approx. 0.56 pounds per gallon

NOTE: Weight of LH2 will vary as much as plus or minus 5%
        due to variations in density.

-end-

www.ingramcontent.com/pod-product-compliance
Lightning Source LLC
Chambersburg PA
CBHW051208200326
41519CB00025B/7044